I0490062

African American Inventors in Science and Technology: Uncovering Hidden Gems

Author: Tracy Webster Jaggi
Publisher: Independently published
Publication date: 2023
ISBN: 9798386992668

Table of Contents

African American Inventors in Robotics and Artificial Intelligence **156**

African American Inventors in Communication Technology **171**

African American Inventors in Nanotechnology **181**

African American Inventors in Environmental Science and Technology **194**

Dedication

To My Dear Father, Raymond B. Webster,

With humble gratitude and inspired determination, I dedicate this book to the one who has always been my source of inspiration and support. Your tireless work in creating your first book and awareness about "African American Firsts in Science and Technology" has left a lasting impact on me and has fuelled my passion for continuing your legacy.

Through your tireless research and dedication, you have shed light on the stories of African Americans who have made tremendous contributions to science and technology and, unfortunately, were unknown to many. Your efforts have highlighted these incredible achievements and inspired future generations to pursue their passions in these fields.

This book pays homage to the many African American pioneers who have paved the way for future generations and inspired and encouraged others to follow in their footsteps. Your work has given us a platform to continue the conversation and celebrate the contributions of African Americans in science and technology.

I am grateful for your guidance and support as I embark on this journey to continue your legacy and bring recognition to the significant contributions of African Americans in science and technology. Thank you for being a role model and always believing in me. It is with great pride that I am here today, honored to be your daughter and carry on your important work.

With love and gratitude,

Your Daughter,
Tracy Webster Jaggi

Forward

Dear Reader,

I invite you to discover innovative stories of African American Inventors in Science and Technology, a tribute to the brave and inventive minds that have changed our world. This book only showcases the achievements of some pioneering African American inventors who broke barriers and made significant contributions to science, technology, engineering, and mathematics. This is only the beginning of who should be recognized.

Discover the lives and careers of these pioneers and embark on a journey of inspiration and motivation. These inventors often had to overcome tremendous adversity to achieve recognition and success, and their stories are a testament to the resilience and determination of the human spirit.

These stories show how difficult it remains for African American inventors to be recognized and appreciated for their work. Despite their challenges, these inventors persevered and contributed to technology and innovation. Their stories of triumphs and hard-fought successes will leave a lasting impression on your mind and soul. By overcoming adversity and discrimination, these heroes paved the way for future generations and left a lasting legacy that continues to influence and inspire us today.

Stories of African American Inventors in Science and Technology provide insight into diversity and inclusion in innovation and how these inventors made a difference in the world. The book also highlights the continuing legacy of these inventors and how their work continues to shape and inspire current and future generations.

What enables African Americans to remain focused and motivated in a society that often ignores their contributions? The answer lies in their unyielding spirit and unwavering determination. The inner qualities that nurtured the souls of African Americans and led them to become leaders and pioneers are captured in this book. The scope of this book covers the pioneering achievements of African Americans over more than a century, from medicine to agriculture, from land surveying to space travel, and from engineering to physics.

"African American Inventors in Science and Technology: Uncovering Hidden Gems" is more than just a list of African American pioneers. It is a tribute to their unique qualities and the intricate connections that formed the basis for African American progress in science and technology. These inventors provide a framework for understanding the diverse attributes that led African Americans to make fundamental cultural advances for themselves, their nation, and the world. These stories make the African American community's richness, beauty, and strength more visible.

This book is a call to action, a prescription for a severe illness, and a celebration of the commitment to excellence and love of humanity that these African American trailblazers embody. Be inspired and captivated by their stories and their innovations' impact on society.

Join us on this unforgettable journey through the lives of some of the most brilliant and influential inventors ever.

Sincerely,
Tracy Webster Jaggi

Introduction

Importance of uncovering the contributions of African American inventors in science and technology

The contributions of African American inventors in science and technology have been overlooked for far too long. Still, it's time to uncover the hidden gems waiting to be discovered. These inventors have made significant contributions to fields such as medical science, computer science, and transportation technology, to name just a few. Acknowledging and celebrating these inventors is vital because they inspire future generations to pursue their dreams and make their mark in the world.

When we learn about the accomplishments of African American inventors, we see that they have overcome tremendous obstacles to achieve great things. Many of these inventors were born into poverty and faced discrimination and racism. Despite these challenges, they persevered and significantly contributed to fields that have transformed our world.

For example, African American inventors have made significant contributions to medical science. Dr. Charles Drew, for instance, developed a method for storing blood plasma, saving countless lives. Meanwhile, Dr. Patricia Bath invented a device that removes cataracts from the eye, improving the vision of millions of people.

In addition to medical science, African American inventors have significantly contributed to computer science, aerospace technology, renewable energy, and communication technology. They have helped shape our world today, and their innovations inspire new generations of inventors and scientists.

By learning about the accomplishments of African American inventors in science and technology, we can better understand our history and the contributions of people from all walks of life. We can also be inspired to pursue our dreams and make our mark. So let's celebrate the hidden gems of African American inventors in science and technology and let their stories inspire us all.

Overview of the book's content and structure

Summary of the book's content and structure

"African American Inventors in Science and Technology: Uncovering Hidden Gems" is a fascinating book that will take you through the history of African American inventors and their groundbreaking contributions to science and technology. This book is specifically designed for children between the ages of 8 and 18 who are curious and eager to learn about important social and environmental issues. It is also an excellent resource for teachers and parents who want to inspire their children to become aware of and engaged in these issues.

The book is divided into eleven chapters, each focusing on a different niche of African American inventors in science and technology. The niches include African American Women Inventors in Science and Technology, African American Inventors in Medical Science, African American Inventors in Computer Science, African American Inventors in Aerospace Technology, African American Inventors in Renewable Energy, African American Inventors in Transportation Technology, African American Inventors in Robotics and Artificial Intelligence, African American Inventors in Communication Technology, African American Inventors in

Nanotechnology, and African American Inventors in Environmental Science and Technology.

Each chapter begins with a brief introduction to the niche and its importance in today's world. The chapters then highlight the achievements of some of the most prominent African American inventors in that field.

The book is filled with fascinating facts and stories about the inventors and their inventions, accompanied by beautiful illustrations. Each chapter also includes a "Did You Know?" section, which provides additional exciting information about the niche, and a "Take Action" section, which encourages readers to get involved in issues related to that niche.

In conclusion, "African American Inventors in Science and Technology: Uncovering Hidden Gems" is an inspiring and educational book that will inspire readers to learn more about the fantastic contributions of African American inventors to science and technology. Whether you are a student, teacher, or parent, this book is an excellent resource for learning about important social and environmental issues and inspiring young people to take action.

*Note: Inventors are often recognized for their groundbreaking contributions to science and technology, and it is not uncommon for them to be featured in multiple book sections. This is because their innovations and inventions often transcend the boundaries of a single field and have applications in different areas.

These inventors possess a diverse range of knowledge and expertise, allowing them to make significant contributions in various fields. Moreover, inventors often build upon existing technologies or combine ideas from different fields to create something new. They are inspired by the work of others and often collaborate across different areas of expertise, leading to the cross-fertilization of ideas and the emergence of new disciplines.

In conclusion, inventors who appear in multiple sections of this book do so because their contributions cross over different fields of science and technology. Their work represents the interconnectedness of other areas of study and highlights the importance of interdisciplinary approaches to problem-solving. Their innovations serve as a testament to the power of collaboration, creativity, and the pursuit of knowledge.

Background Information

STEM stands for Science, Technology, Engineering, and Mathematics

Introduction

STEM education is essential for all communities, including African Americans. STEM stands for Science, Technology, Engineering, and Mathematics, and it is a set of fields crucial for understanding the world around us, solving problems, and developing new technologies. Here, we'll explore why STEM education is essential, how it can benefit African Americans, and how it has played a role in the accomplishments of African American inventors.

Why STEM Education is Important

STEM education is vital for many reasons. First, STEM careers are in high demand and tend to pay well. By pursuing a STEM education, African Americans can access various career opportunities, from computer programming to biomedical research. STEM jobs are also projected to grow faster than non-STEM jobs, so investing in a STEM education can increase job security and financial stability.

Second, STEM education teaches critical thinking, problem-solving, and creativity. These skills are essential for all fields, especially STEM, where innovation and creativity are crucial to developing new technologies and solving complex problems.

Third, many of society's biggest challenges, such as climate change, healthcare disparities, and inequality, require STEM solutions. By pursuing STEM education and careers, African Americans can help address these challenges and positively impact their communities and the world.

Finally, there still needs to be more diversity in STEM fields. Encouraging more African Americans to pursue STEM can increase representation and diversity, leading to new perspectives, ideas, and more significant innovation.

In addition, STEM education can help address issues of inequality and discrimination. For example, African American women are underrepresented in STEM fields, but pursuing STEM education and careers can challenge stereotypes and pave the way for future generations.

The Role of STEM in the Accomplishments of African American Inventors

STEM has played a significant role in the accomplishments of African American inventors throughout history. Many African American inventors have significantly contributed to science, technology, engineering, and mathematics. For example, George Washington Carver was a botanist and agricultural scientist who developed new uses for crops like peanuts and sweet potatoes. Katherine Johnson was a mathematician who helped NASA send astronauts to the moon.

By learning about these and other African American inventors, young readers can see how STEM can be used to make the world a better place. They can also be inspired to pursue their STEM interests and contribute to science, technology, engineering, and math.

Conclusion

STEM education is essential for the African American community. It can lead to more excellent career opportunities, innovation and creativity, solutions to societal challenges, and increased representation and diversity in STEM fields. Investing in a STEM education and encouraging more African Americans to pursue STEM careers can create a brighter future for everyone.

African American Inventors in Science and Technology

Despite facing substantial barriers and discrimination, African Americans have contributed significantly to science and technology for centuries. In this chapter, we explore the stories of inventors such as Patricia Bath, who revolutionized eye surgery with her invention of the Laserphaco Probe, and Dr. George Washington Carver, who revolutionized agriculture with his work on crop rotation and soil conservation. These inventors, and many others like them, have helped to shape the world we live in today, and their contributions continue to inspire future generations of inventors.

Benjamin Banneker

Benjamin Banneker: A Pioneer in Science and Mathematics

Benjamin Banneker was an African American inventor, mathematician, and astronomer who lived in the late 1700s. He is often considered one of the first African American scientists in history. Despite being born into slavery, Banneker was a self-taught genius who contributed significantly to astronomy, mathematics, and engineering.

Banneker's most notable achievement was his creation of the first clock made entirely from wood. He built this clock in 1753, at the age of 21, and it was so precise that it continued to work for more than 50 years. Banneker's clock was a remarkable engineering feat, especially considering he had no formal training in the field.

In addition to his work as a clockmaker, Banneker was also an accomplished astronomer. Using his calculations and observations, he was the first to predict a solar eclipse in 1789. He also published a series of almanacs that included information about the positions of the stars and planets and important dates and events.

Despite facing discrimination and prejudice throughout his life, Banneker continued to pursue his passion for science and mathematics. He used his knowledge to challenge the prevailing beliefs of his time, and his work paved the way for future generations of African American scientists and inventors.

Banneker's legacy inspires young people today, especially those interested in pursuing careers in science, technology, engineering, and mathematics (STEM). His story reminds us that with hard work, determination, and a thirst for knowledge, anyone can achieve great things, regardless of background or circumstances.

In conclusion, Benjamin Banneker was a pioneer in science and mathematics whose groundbreaking work paved the way for future generations of African American inventors and scientists. His legacy is a testament to the power of education and perseverance, and his story inspires all those seeking to make a difference in the world.

George Washington Carver

George Washington Carver

George Washington Carver was an African American inventor, scientist, and educator who significantly contributed to agriculture and botany. He was born into slavery in Missouri in 1864 and faced many challenges throughout his life, but he never gave up on his dreams of learning and making a difference in the world.

Carver is best known for his work with peanuts, but he also researched and developed new uses for other crops, such as sweet potatoes, soybeans, and pecans. He believed these crops could help improve the lives of farmers and their families, and he worked tirelessly to find new and innovative ways to use them.

One of Carver's most influential inventions was peanut butter. He discovered that by roasting and grinding peanuts, he could create a nutritious and tasty spread that could be used in various ways. Peanut butter soon became a popular food item and helped to make peanuts a valuable crop for farmers.

Carver also developed new uses for cotton, the main crop grown in the southern United States at the time. He found ways to use cottonseed oil in cooking and as a lubricant for machinery, and he even developed a type of paper made from cotton waste.

In addition to his scientific work, Carver was also a dedicated educator. He taught at the Tuskegee Institute in Alabama for many years and inspired countless students to pursue their dreams and make a difference in the world.

Carver's legacy inspires people today, and his work has impacted agriculture and botany. He is a shining example of the power of perseverance, dedication, and innovation, and his story can inspire us all to make a difference in the world.

Granville T. Woods

Granville T. Woods

Granville T. Woods was an African American inventor who significantly contributed to electrical engineering. Born in 1856, Woods grew up in Ohio and attended school until age ten, when he had to leave to work and support his family. Despite this setback, he continued to educate himself by reading and studying independently.

Woods began his career as an apprentice in a machine shop, where he learned about steam engines and locomotives. He soon realized he had a talent for inventing and began developing his ideas. His first invention was a steam boiler furnace that could burn coal more efficiently, which earned him a patent. Over the years, he went on to invent many more devices, including a telegraph system that allowed trains to communicate with each other, a device that improved the safety of elevator systems, and an automatic air brake for trains.

One of Woods' most significant inventions was the multiplex telegraph, which allowed messages to be sent simultaneously over a single wire. This invention revolutionized communication and paved the way for developing the telephone and other modern communication systems.

Despite facing discrimination and racism throughout his career, Woods continued to invent and improve upon existing technology. He received over 60 patents in his lifetime and became known as the "Black Edison" for his contributions to electrical engineering.

Granville T. Woods' legacy serves as an inspiration to all young people who are interested in science and technology. His determination and perseverance in adversity show that anyone can achieve great things if they work hard and believe in themselves. By learning about Woods and other African American inventors, we can better understand the significant contributions of people of all races and backgrounds to science and technology.

Jan Ernst Matzeliger

Jan Ernst Matzeliger: A Revolutionary African American Inventor

Jan Ernst Matzeliger was a revolutionary African American inventor who changed the world of shoemaking with his lasting machine invention. Born in Paramaribo, Surinam, in 1852, Matzeliger grew up with a fascination for mechanics and engineering. He moved to the United States in 1871 and settled in Lynn, Massachusetts. He worked in a shoe factory and held several jobs, including operating a sole sewing machine, running a heel burnisher, and repairing machinery.

Despite facing discrimination and prejudice as an African American inventor working in a primarily white, male-dominated industry, Matzeliger was determined to make a difference. He spent years developing his lasting machine invention, patented in 1883 and considered the most extraordinary step in the shoemaking industry. His machine could last shape the upper leather over the last and attach it to the bottom of the shoe with great speed and efficiency, significantly increasing productivity and efficiency in the industry.

Matzeliger was an inventor and a leader in the African American community. He worked to promote the advancement of African Americans in engineering and mechanics and advocated for education. His contributions to the African American community were immeasurable, and he served as a role model for many.

In addition to his lasting machine invention, Matzeliger also invented the machine that trimmed and shaped the edges of the soles and the machine that cut the tops of the shoes. He also made several other improvements to the shoemaking process, including a device that attached the soles to the uppers and a machine that perforated the soles to attach the heels. Matzeliger's contributions to the shoe industry were not limited to one invention but rather a series of innovations that significantly improved the efficiency and productivity of the shoemaking process.

Despite facing many challenges, including discrimination and prejudice, Matzeliger never gave up on his dream of making a difference in the world. His legacy is one of perseverance, determination, and community contributions, inspiring future generations of inventors and engineers, particularly those from underrepresented groups. Today, Matzeliger's lasting machine is still celebrated as an essential milestone in the history of the shoe industry, and his story continues to inspire and influence people worldwide.

Percy Julian

Percy Julian: The Man Who Made Medicine More Accessible

Percy Julian was a man ahead of his time. He faced numerous obstacles during his lifetime, but his determination and perseverance made him a trailblazer in medical science. Julian's most significant contribution to medicine was developing a synthetic version of cortisone. This hormone is used to treat inflammation and other medical conditions. Before Julian's invention, cortisone was only available in small quantities from animal glands, which made it expensive and difficult to obtain. Julian's synthetic cortisone was much more affordable and accessible to people who needed it.

Julian's contributions to science were not limited to the development of synthetic cortisone. He also made important discoveries in the field of plant chemistry. He developed a process for extracting soy protein to create new products, such as soy milk and soy ice cream. He also discovered a way to remove sterols from soybean oil, which led to the development of a new birth control pill.

Julian's work in the field of medical science and plant chemistry had a significant impact on the world. He made medicine more accessible to people who needed it, and his discoveries in plant chemistry led to the development of new products that improved people's lives. Julian's contributions to science and medicine were groundbreaking but faced many challenges.

As an African American scientist, Julian was often discriminated against and had to work twice as hard to achieve success. He faced discrimination from some of his classmates and professors while studying at DePauw University. However, he persevered and earned his Ph.D. from the University of Vienna in Austria. When he returned to the United States, he became a research chemist and worked for several companies, including Glidden Company and Soya Products Company.

Julian's success as a scientist resulted from his intelligence and hard work, and ability to overcome obstacles and push through challenges. He never gave up on his dreams and continued to make important discoveries throughout his career. His legacy continues to inspire young scientists today.

In conclusion, Percy Julian was a man who made medicine more accessible. His contributions to medical science and plant chemistry were groundbreaking, and his legacy inspires young scientists today. Despite facing discrimination and other obstacles, Julian never gave up on his dreams. His determination and perseverance made him a trailblazer in medical science and an inspiration to all who strive to make a difference in the world.

Lonnie G. Johnson

Lonnie G. Johnson

Lonnie G. Johnson is an African American inventor who has significantly impacted the world. He was born on October 6, 1949, in Mobile, Alabama, and grew up when segregation was legal in the United States. Despite facing many challenges and obstacles, Johnson pursued his passion for science and technology and became one of the most successful inventors of his time.

One of Johnson's most famous inventions is the Super Soaker, a powerful water gun that has provided countless hours of fun for children and adults worldwide. Johnson devised the idea for the Super Soaker while working on a new cooling system for NASA's space program. He noticed that the high-pressure stream of water from the nozzle could also be used as a toy, and he decided to develop it further. The Super Soaker has since become one of the most popular toys ever, selling over 200 million units and generating billions of dollars in revenue.

Besides the Super Soaker, Johnson has also invented other essential technologies that have contributed to society. He holds over 120 patents for his inventions, including a thermoelectric energy converter, a ceramic battery, and a heat pump that can be used for air conditioning and refrigeration. Johnson's inventions have revolutionized various industries, and his commitment to innovation and creativity has inspired many young people to pursue careers in science and technology.

Despite facing discrimination and prejudice throughout his life, Johnson has remained steadfast in his pursuit of excellence. He received numerous awards and honors, including the National Medal of Technology and Innovation in 2015, and he continues to work on new inventions that can positively impact the world.

Lonnie G. Johnson's story is a testament to the power of perseverance, creativity, and innovation. His inventions have brought joy and excitement to millions worldwide, and his legacy will continue to inspire future generations of inventors and innovators.

Mark E. Dean

Mark E. Dean

Mark Dean is an African American inventor who has made groundbreaking contributions to the world of computer science. Born in Jefferson City, Tennessee, in 1957, Dean was interested in technology from a young age. He studied electrical engineering at the University of Tennessee, earning his Bachelor's, Master's, and Ph.D. degrees.

One of Dean's most significant contributions to computer science was his work on developing the IBM personal computer. He led a team of engineers that designed the hardware and software for the original model, which was introduced in 1981. The IBM PC was a game-changer in computing, making computers more accessible and affordable for everyday people.

Dean also invented the first color PC monitor and gigahertz chip, which is still used in many devices today. His work has been recognized with numerous awards, including induction into the National Inventors Hall of Fame in 1997.

In addition to his computer science work, Dean strongly advocates diversity in STEM fields. He has spoken about the importance of encouraging underrepresented groups, including people of color and women, to pursue careers in science and technology.

Dean's legacy serves as an inspiration to young people who are interested in technology and engineering. His groundbreaking work has paved the way for many of the devices and technologies we use today, and his advocacy for diversity in STEM fields is an important reminder that everyone has the potential to make a difference.

As children and teens interested in science and technology, learning about the contributions of African American inventors like Mark Dean is vital. By studying their work and learning from their experiences, we can better understand the diverse perspectives and skills needed to solve our world's complex problems.

Shirley Ann Jackson

Shirley Ann Jackson: A Pioneer in Physics and Technology

Shirley Ann Jackson is a remarkable African American inventor who has significantly contributed to physics and technology. She was born on August 5, 1946, in Washington, D.C., and was the first African American woman to earn a Ph.D. from the Massachusetts Institute of Technology (MIT).

Jackson's research focused on theoretical physics and particle physics, which involved studying the behavior of subatomic particles. She also worked on developing new technologies, including the touch-tone telephone, fiber optic cables, and solar cells. Her work has been instrumental in advancing telecommunications, energy, and environmental science.

In 1995, Jackson became the first African American woman to lead a major research university when she was appointed president of Rensselaer Polytechnic Institute (RPI) in Troy, New York. She used her position to promote STEM education and increase diversity in the field.

Jackson has received many awards and honors for her contributions to science and education, including the National Medal of Science, the highest honor given to scientists in the United States. She has also been inducted into the National Women's Hall of Fame and the American Academy of Arts and Sciences.

Jackson's story is a testament to the importance of education and perseverance. Despite facing discrimination and obstacles, she pursued her passion for science and technology and made groundbreaking discoveries that have changed our world. She inspires young people, especially girls and minorities, who may face similar challenges in pursuing their dreams.

As we continue to face global challenges such as climate change and energy conservation, it is vital to have diverse voices and perspectives in the scientific community. Shirley Ann Jackson's legacy serves as a reminder of the importance of inclusion and diversity in STEM fields and the impact one person can have on the world.

James West

James West

James West is an African American inventor who has made significant contributions to the field of acoustics. Born in 1931 in Virginia, West grew up in a segregated society where opportunities for African Americans were limited. However, he was determined to succeed and pursued his passion for science and technology.

West attended Temple University in Philadelphia, where he earned a degree in physics. After graduation, he worked at Bell Laboratories, a research center for the communications industry. Here, West made his most significant invention – the electret microphone.

The electret microphone is a small device that converts sound waves into electrical signals. It is used in various applications, including telephones, hearing aids, and recording equipment. West's invention revolutionized the field of acoustics and paved the way for many other innovations.

West's work at Bell Laboratories also included research on other aspects of acoustics, such as noise reduction and acoustic imaging. He has received numerous awards and honors for his contributions to the field, including the National Medal of Technology and Innovation.

In addition to his scientific work, West has advocated for diversity and inclusion in STEM fields. He has mentored many young scientists and encouraged them to pursue their dreams, regardless of their background.

James West's story inspires all young people who aspire to make a difference in the world through science and technology. Despite facing obstacles and discrimination, he persevered and made significant contributions to his field. His work has improved the lives of countless people, and his legacy will continue to inspire future generations of inventors and innovators.

Patricia Bath

Patricia Bath

Patricia Bath was an African American inventor who revolutionized eye surgery with her invention of the Laserphaco Probe. Born in Harlem, New York, in 1942, Bath grew up when African Americans faced significant discrimination and limited opportunities. Despite these challenges, Bath was determined to pursue her passion for science and technology.

After earning her bachelor's degree in physics from Hunter College, Bath attended medical school at Howard University. She became the first African American woman to complete a residency in ophthalmology at New

York University. During her time as a resident, Bath became interested in developing a new tool for cataract surgery.

At the time, cataract surgery was a painful and invasive procedure that often resulted in complications such as infection and permanent vision loss. Bath believed there had to be a better way to perform the surgery. She began working on a new device using a laser to break up the cataract and a suction tool to remove it from the eye.

Bath's invention, the Laserphaco Probe, was a game-changer for cataract surgery, and it made the procedure faster, safer, and less painful for patients. The Laserphaco Probe was also more precise than previous methods, resulting in better patient outcomes.

Bath's invention was granted a patent in 1988, making her the first African American woman to receive a medical patent. She went on to use her invention to help restore the sight of people around the world, particularly those in underserved communities.

In addition to her work in medicine, Bath advocated for women and minorities in science and technology. She founded the American Institute for the Prevention of Blindness to provide vision care and education to underserved communities. Bath passed away in 2019, but her legacy as a pioneer in medicine and a champion for social justice lives on.

Patricia Bath's story is an inspiration to young people who are interested in science and technology. Her determination and innovation changed ophthalmology and helped improve countless people's lives. By learning about Patricia Bath and other African American inventors in science and technology, we can see the significant contributions that people from all backgrounds can make to our world.

Otis Boykin

Otis Boykin: The Inventor Who Changed the World

Have you ever heard of Otis Boykin? If you still need to, you're missing out on one of the most influential African American inventors in science and technology. Otis Boykin was an inventor who changed the world with his incredible inventions, and his work continues to impact our lives today.

Born in 1920 in Texas, Boykin grew up in a world where African Americans faced discrimination and segregation. Despite these challenges, Boykin was determined to make a difference. He studied at Fisk University and the

Illinois Institute of Technology, where he earned a degree in electrical engineering.

Boykin's first invention was a device called the "variable resistor." This device was used to control the flow of electricity in electronic devices and was a breakthrough in electronics. Boykin's invention was so important that it was used in many devices, including pacemakers, computers, and even guided missiles.

But Boykin didn't stop there. He went on to invent many other devices that changed the world. One of his most important inventions was the "control unit" for the pacemaker. This device helped regulate the heartbeat of people with heart problems, saving many lives.

Boykin also invented the "resistor-capacitor" network used in televisions and radios. This device helped to improve the quality of sound and pictures in these devices, and it was a breakthrough in the world of electronics.

Throughout his life, Boykin received many awards and honors for his inventions. He was inducted into the National Inventors Hall of Fame in 1990, and his work continues to inspire young people worldwide.

Otis Boykin was an incredible inventor who changed the world with his inventions. He faced many challenges in life but never gave up on his dream of making a difference in the world. If you're interested in science and technology, you should learn more about Otis Boykin and his amazing inventions.

Sandra K. Johnson

Sandra K. Johnson: The Fearless Innovator
Shattering Glass Ceilings

Meet Sandra K. Johnson, a true trailblazer who shattered the glass ceiling in the technology industry. Johnson was born in the 1960s and grew up in a world where white men dominated technology. Despite this, she was driven by her passion for technology and unylelding determination to make her mark in a field that had long excluded women and people of color.

Johnson's journey began at Southern University, where she earned her bachelor's degree in electrical engineering with honors. She continued her education at Stanford University, receiving her master's degree in electrical engineering. Then she earned her Ph.D. in electrical engineering from Rice University, becoming the first African American female to do so.

Johnson made history in her twenty-four-year tenure at IBM and broke down barriers. She was responsible for groundbreaking innovations in the technology industry, including developing the base machine for "Deep Blue," IBM's world-famous chess machine. Johnson also acquired twenty patents and several other patent applications pending, researched various high-end computer-related areas, and held several high-ranking positions, including Linux Performance Architect and managing the Linux Performance team.

But Johnson's impact goes far beyond her professional accomplishments. She has actively participated in various non-profit organizations and community programs, advocating for diversity and inclusion in the tech industry and mentoring young women and people of color. She is also a devoted wife and mother who has always balanced her professional and personal life, prioritizing her family while achieving great success in her career.

Johnson's legacy serves as an inspiration for all young people, especially those from underrepresented communities. Her unwavering commitment to excellence and her courage in the face of adversity testify to the power of perseverance and the importance of breaking down barriers. Sandra K. Johnson is a true pioneer, a leader, and a role model, and her impact on the technology industry will continue to inspire generations to come.

African American Women Inventors in Science and Technology

Despite facing significant barriers and discrimination, African American women have made remarkable contributions to science and technology. In this chapter, we explore the stories of women such as Dr. Gladys West, who invented the technology that underlies GPS navigation, and Marie Van Brittan Brown, who created the precursor to the modern home security system. These women, and many others like them, have defied the odds to make their mark on history, and their contributions continue to inspire future generations of inventors.

Mary Kenner

Mary Kenner: A Trailblazing Inventor Who Changed the Game

Mary Kenner was a prolific inventor and entrepreneur who significantly impacted the world of feminine hygiene. Born in Monroe, North Carolina, in 1912, she was the daughter of a successful inventor and entrepreneur, and from a young age, she was inspired by her father's work and his passion for innovation.

As a young woman, Kenner noticed that no effective menstrual hygiene solutions were available to women. She saw that the standard answers at the time - cloth pads and tampons -

were uncomfortable and often leaked, causing embarrassment and discomfort for women.

So, she set to work inventing a new kind of maxi pad that would be more comfortable, absorbent, and effective. She spent years researching and testing different materials and designs, and in 1956, she finally patented her invention, the Sanitary Belt.

The Sanitary Belt was a game-changer. It was a belt with adjustable straps that held a disposable pad in place, allowing women to move more freely and comfortably during their menstrual cycle. It was a huge success, and Kenner went on to create several more inventions in the field of feminine hygiene, including a toilet tissue holder and a moisture-resistant shower cap.

Despite her groundbreaking work, Kenner faced significant challenges as a black woman in a male-dominated field. She struggled to secure funding for her inventions and faced discrimination and sexism. However, she persevered, and her stories profoundly impacted women's lives worldwide.

Kenner's legacy is one of innovation, determination, and perseverance. She was a true trailblazer who paved the way for countless other inventors, and her contributions to feminine hygiene have impacted women's lives. Today, we honor her memory and celebrate her achievements, and we continue to be inspired by her example of creativity, ingenuity, and courage in the face of adversity.

Alice Parker

Alice Parker

Alice Parker was an African American inventor who revolutionized how we heat our homes. Born in 1895 in Morristown, New Jersey, Parker was the daughter of working-class parents who valued education. Her father was a tailor, and her mother was a homemaker who instilled a love of learning in her children. Parker was an excellent student, and she excelled in math and science.

In the early 1900s, most homes were heated with coal or wood stoves, which were messy, inefficient, and dangerous. Parker saw an opportunity to improve this technology and began working on a new heating system that would use natural gas. She spent years conducting experiments in her home, testing different designs and materials. Finally, in 1919, she patented her invention, which became known as the gas furnace.

Parker's gas furnace was a game-changer. It was safer, cleaner, and more efficient than traditional heating systems. It used natural gas, which was abundant and relatively inexpensive, and it could heat multiple rooms at once. Parker's invention allowed people to live comfortably in colder climates and paved the way for modern heating systems.

Despite her groundbreaking invention, Parker faced many obstacles as a black woman in a male-dominated field. She struggled to find investors and faced discrimination from companies and government agencies. However, she persevered and continued to innovate throughout her life. She patented several other inventions, including a new kitchen stove and a device for cleaning chimneys.

Today, Alice Parker's legacy lives on in the millions of homes and buildings that use natural gas heating. Her invention has significantly impacted the environment, reducing traditional heating systems' pollution and greenhouse gas emissions. Parker's story is a testament to the power of innovation and determination, inspiring future   generations   of   inventors.

Marie Van Brittan Brown

Marie Van Brittan Brown

Marie Van Brittan Brown was an African American inventor who lived in Queens, New York, during the 1960s. She was a nurse who worked in a hospital and was concerned about the safety of her home. At that time, crime rates in New York were high, and Brown wanted a way to protect herself and her family.

Brown's solution was to develop a home security system. She invented a closed-circuit television system (CCTV) that could be installed in homes to monitor potential intruders. The design included a camera that could be mounted outside the front door and a monitor to display the footage inside the house.

Brown's invention was unique because it included a two-way communication system, which meant the homeowner could see the person at the door and speak to them through a microphone. The invention was also equipped with a remote control to unlock the door from a distance, allowing the homeowner to let in a trusted visitor without opening the door.

Brown's revolutionary invention paved the way for modern-day home security systems. Her invention was patented in 1969 and was the first of its kind. Brown's design was important not only for home security but also for advancing technology in general.

Marie Van Brittan Brown's invention was a game-changer in home security. Her creativity and innovative spirit led to a safer future for homeowners everywhere. Today, her invention is still used as a blueprint for modern-day home security systems. Brown's legacy inspires young people interested in science and technology, especially African American girls. Her story reminds us that anyone can be an inventor, regardless of background or circumstances.

Dr. Gladys West

Dr. Gladys West

Dr. Gladys West is a remarkable African American inventor who has significantly contributed to science and technology. Born in 1930 in Virginia, Dr. West grew up when opportunities for women and people of color were limited. However, she was determined to pursue her passion for mathematics and science.

Dr. West studied at Virginia State College, where she excelled in her studies and earned a degree in mathematics. She earned a master's in mathematics from the University of Virginia and later received a fellowship to study at the Naval Surface Weapons Center in Virginia.

At the Naval Surface Weapons Center, Dr. West began her groundbreaking work in geodesy. Geodesy is the science of measuring the shape and size of the Earth, and it is essential for many critical applications, such as navigation and mapping.

Dr. West was one of the first people to use complex mathematical equations to model the shape of the Earth, taking into account its irregularities and variations in gravity. Her work helped to improve the accuracy of GPS systems, which are now used in everything from cars to airplanes to smartphones.

Despite facing discrimination and barriers as a woman and a person of color, Dr. West persevered and made significant contributions to science and technology. She was inducted into the Air Force Space and Missile Pioneers Hall of Fame in 2018, and her legacy continues to inspire young people today.

Dr. Gladys West's story is a testament to the power of determination and perseverance, and it inspires all young people who dream of making a difference in the world.

Dr. Shirley Ann Jackson

Dr. Shirley Ann Jackson

Dr. Shirley Ann Jackson is a brilliant African American inventor in the field of physics. She was born on August 5, 1946, in Washington, D.C., and grew up in a family that valued education. Her parents encouraged her to pursue her passion for science, and she excelled in school, earning a scholarship to attend the Massachusetts Institute of Technology (MIT).

At MIT, Dr. Jackson studied physics and became the first African American woman to earn a Ph.D. from the university. Her research focused on theoretical physics, and she made groundbreaking discoveries in the field. Her work led to the development of new technologies, including the touch-tone telephone, fiber-optic cables, and the portable fax machine.

Dr. Jackson's contributions to science and technology have been recognized with numerous awards and honors. In 2014, she was awarded the National Medal of Science, the highest honor given to scientists in the United States. She is also a member of the National Academy of Sciences, the American Academy of Arts and Sciences, and the American Philosophical Society.

Dr. Jackson has also used her position in the scientific community to advocate for diversity and inclusion. She has worked to increase the number of women and minorities in the science and technology fields. She has served on several advisory boards and committees promoting STEM diversity.

Dr. Shirley Ann Jackson is a true inspiration to young people interested in science and technology. Her groundbreaking research and advocacy for diversity and inclusion have significantly impacted the field of physics and beyond. Her story reminds us that anyone can make a difference with hard work, perseverance, and commitment to improving the world.

Dr. Patricia Bath

Dr. Patricia Bath

Dr. Patricia Bath was an African American medical science inventor who significantly contributed to the field of ophthalmology. She was born in Harlem, New York, on November 4, 1942, and grew up in a family that valued education and hard work.

Dr. Bath was the first African American woman doctor to receive a patent for a medical invention. Her invention was called the Laserphaco Probe, a revolutionary device that used laser technology to remove cataracts from the eye. This device was a game-changer in the medical field because it allowed for quicker and more accurate removal of cataracts, a common eye condition that can lead to blindness.

Dr. Bath's interest in ophthalmology began when she was a teenager and worked as a summer intern at a Harlem hospital. She noticed that many patients had preventable blindness due to poor access to proper eye care. This experience inspired her to pursue a medical career and work towards solving this problem.

Throughout her career, Dr. Bath was a leader and advocated for diversity and inclusion in the medical field. She co-founded the American Institute for the Prevention of Blindness and was a founding member of the International Society for Eye Research. She was the first female African American ophthalmologist at UCLA Medical Center and the Jules Stein Eye Institute.

Dr. Bath's work and achievements have been recognized with numerous awards and honors, including induction into the National Inventors Hall of Fame in 2005. Her legacy continues to inspire and encourage young people, particularly girls and students of color, to pursue careers in science, technology, engineering, and mathematics (STEM) fields.

Dr. Patricia Bath was a pioneering African American medical science inventor who significantly contributed to the field of ophthalmology. Her Laserphaco Probe invention has helped countless people regain sight and lead better lives worldwide. Her work also serves as an inspiration for young people to pursue their dreams and make a positive impact on the world.

Dr. Mae Jemison

Dr. Mae Jemison: The First African American Woman in Space

Dr. Mae Jemison is a name you may not have heard before, but she is essential in science and technology. Dr. Jemison is the first African American woman to travel into space and has accomplished many other amazing things.

Dr. Jemison was born in Alabama in 1956 and grew up in Chicago with parents who encouraged her to pursue her interests and dreams. She was always interested in science and space and worked hard in school to become a doctor. After earning a degree in chemical engineering, she attended medical school and eventually became a doctor.

But Dr. Jemison didn't stop there. She applied to become an astronaut and was accepted into the NASA space program in 1987. In 1992, she traveled aboard the Space Shuttle Endeavour, becoming the first African American woman in space. During her eight-day mission, she conducted experiments and studies on the effects of weightlessness on the human body.

After leaving NASA, Dr. Jemison continued to work on meaningful projects related to science and technology. She founded the Jemison Group, which focuses on developing new technologies to solve problems in developing countries. She has also been a professor, a public speaker, and a strong advocate for the importance of science education.

Dr. Jemison's accomplishments are truly inspiring, and she serves as a role model for young people interested in science and technology. Her story shows that anyone can achieve their dreams and make a difference in the world with hard work, determination, and a passion for learning.

Dr. Mae Jemison is a hidden gem in the world of African American Inventors in Aerospace Technology. Her contributions to science and technology have been groundbreaking and inspiring. Her story is important to share with children and teens curious and eager to learn about important social and environmental issues. By learning about Dr. Jemison and other African American inventors in science and technology, young people can be inspired to pursue their interests and positively impact the world.

Ursula M. Burns

Ursula M. Burns: A Pioneer and Leader Who Broke the Glass Ceiling

Ursula M. Burns is a name that has become synonymous with perseverance, resilience, and determination. She is a woman who has defied the odds and broken barriers to become one of the most influential figures in the business world. Her journey is a testament to the power of hard work and dedication, and her story inspires people from all walks of life.

Born in New York City in 1958, Burns grew up in a tough neighborhood where poverty and discrimination were common. Despite these challenges, she was determined to succeed and pursue her passion for engineering. She studied mechanical engineering at the Polytechnic Institute of New York and later earned her M.S. in mechanical engineering from Columbia University.

Burns began her career at Xerox in 1980 as a mechanical engineering intern. Over the years, she worked her way up through the ranks, receiving promotions in both engineering and management. She faced discrimination and bias as a woman and an African American, but she never let these obstacles stop her.

In 2009, Burns made history by becoming the first African American female CEO to lead a Fortune 500 company. She led Xerox through significant change and transformation, focusing on growth areas like services and software. Under her leadership, the company made substantial acquisitions, such as Affiliated Computer Services (ACS), and remained competitive in a rapidly changing industry.

Throughout her career, Burns has advocated for diversity and inclusion in the workplace. She has encouraged other women and minorities to pursue careers in STEM fields and has served as a mentor and role model for many young people.

In addition to her professional achievements, Burns is also known for her philanthropic work. She serves on the boards of several organizations, including the American Museum of Natural History and the Girl Scouts of the USA. She has also been honored with numerous awards and honors for her contributions to business and society, including being named one of Forbes' 100 Most Powerful Women.

Ursula M. Burns is a true trailblazer who has inspired countless people with her incredible journey. She faced discrimination and obstacles throughout her career but never let them stop her from achieving her goals. Her leadership and advocacy have impacted the business world, and her story is a testament to the power of hard work and perseverance.

African American Inventors in Medical Science

The creative and innovative work of African American inventors has transformed the field of medical science. In this chapter, we explore the stories of inventors such as Dr. Charles Drew, whose research on blood plasma has saved countless lives. These inventors have developed new treatments and devices, challenged the status quo, and paved the way for a more diverse and inclusive medical profession.

Dr. Charles Drew

Dr. Charles Drew: A Pioneer in Blood Plasma Research

Dr. Charles Drew was a brilliant medical researcher who made groundbreaking contributions to the field of blood transfusion. Born in 1904 in Washington, D.C., he grew up in a family of educators and excelled in academics from a young age. Drew attended Amherst College, where he was an outstanding athlete and scholar. After completing his undergraduate studies, he pursued a medical degree at McGill University in Montreal.

In the 1930s, Drew began researching methods for preserving blood plasma, the clear,

75

yellowish fluid that remains after removing red and white blood cells. At the time, blood transfusions were a risky and often fatal procedure, as blood could only be stored briefly before it spoiled. Drew's research led to the development of a technique for separating plasma from blood, which could then be stored for extended periods. This breakthrough allowed blood to be shipped and stored in blood banks for emergencies, saving countless lives.

During World War II, Drew was appointed the director of the blood bank of the American Red Cross. He worked tirelessly to ensure that blood was collected and distributed efficiently to the military, and he also fought against racial discrimination in blood donation. Drew was outraged when the army began segregating blood donations based on race, and he resigned from his position in protest.

Tragically, Drew died in a car accident in 1950 at the age of 45. His legacy, however, lives on as a pioneer in blood plasma research and a champion of racial equality in the medical field. Today, blood banks worldwide continue to use his methods for preserving blood plasma and saving lives.

In conclusion, Charles Drew was an extraordinary African American inventor who significantly contributed to medical science. His research and innovations have saved countless lives, and his dedication to racial equality in the medical field inspires us all. By learning about his achievements, we can gain a deeper understanding of the critical role that African American inventors have played in shaping our world.

Dr. Percy Julian

Dr. Percy Julian

Dr. Percy Julian was one of the most influential African American inventors in chemistry. He was born in Alabama in 1899, and despite facing many challenges because of his race, he became one of the most respected scientists of his time.

Dr. Julian studied at DePauw University in Indiana, where he earned his Bachelor's degree in chemistry. He then earned a Ph.D. in organic chemistry from the University of Vienna in Austria. After completing his education, Dr. Julian worked for several companies, including the Glidden

Company, where he made some of his most important discoveries.

One of Dr. Julian's most famous inventions was a way to make cortisone, a medicine that helps reduce inflammation. Before Dr. Julian's invention, cortisone was expensive and challenging to produce. But thanks to his work, it became much easier and cheaper to make, which made it more widely available to people who needed it.

Dr. Julian also discovered a way to extract soy protein, which is used in many foods today, including tofu and meat substitutes. His work with soy helped make it an essential crop in the United States, and he is often called the "soybean chemist" because of his contributions to the field.

But despite his many accomplishments, Dr. Julian faced discrimination throughout his life. He was often denied opportunities because of his race, and he had to work much harder than his white colleagues to achieve success. But he never gave up and continued to push forward, breaking down barriers and paving the way for future generations of African American scientists.

Today, Dr. Julian is remembered as a hero and a pioneer in chemistry. His inventions and discoveries have had a lasting impact on our world, and he inspires young people everywhere who dream of making a difference in science and technology.

Dr. Samuel Kountz

Dr. Samuel Kountz: A Pioneer in Organ Transplantation

Dr. Samuel Kountz was a pioneering African American inventor in the field of medical science. He dedicated his life to developing techniques for organ transplantation, which have saved countless lives worldwide.

Born in 1930 in Arkansas, Kountz grew up in poverty and faced many obstacles. Despite these challenges, he was determined to pursue a career in medicine. He earned his medical degree from the University of Arkansas in 1958 and completed his residency at Stanford University.

Kountz's interest in organ transplantation began in the early 1960s when he began working with Dr. Roy Cohn, a renowned transplant surgeon. Together, they developed new techniques for performing kidney transplants, including immunosuppressive drugs to prevent rejection.

In 1963, Kountz performed the first successful kidney transplant between two people who were not related. The patient, a 22-year-old woman, lived for 22 years after the transplant. This breakthrough paved the way for many more successful transplants, and Kountz became known as a pioneer in the field.

Throughout his career, Kountz continued to develop new organ transplantation techniques and train other surgeons in his methods. He also worked to address the shortage of donor organs by advocating for organ donation and transplantation.

Despite facing discrimination and racism throughout his life, Kountz never gave up on his dreams or his commitment to helping others. His legacy lives on in the countless lives saved through his work and the many medical professionals who continue to use his techniques today.

Dr. Samuel Kountz is an inspiration to all of us, and his legacy reminds us of the importance of perseverance, dedication, and innovation in the face   of   adversity.

Dr. Alexa Canady

Dr. Alexa Canady: Breaking Barriers in Medicine

Dr. Alexa Canady is an African American inventor in medical science who broke barriers and paved the way for future generations of women and minorities in the field. She was the first African American woman to become a neurosurgeon in the United States.

Born in Lansing, Michigan, in 1950, Dr. Canady was raised in a family of educators who instilled in her the importance of education and hard work. She excelled in school and earned a zoology degree at the University of Michigan. From there, she attended medical school at the same university and completed her residency in neurosurgery at the University of Minnesota.

Dr. Canady faced discrimination and sexism throughout her career but persevered and became a highly respected neurosurgeon. She specialized in pediatric neurosurgery and was known for her compassionate and patient-focused approach to medicine.

Dr. Canady's contributions to medicine extend beyond her groundbreaking achievements as the first African American woman neurosurgeon. She also pioneered new techniques in neurosurgery and published numerous articles and book chapters on the subject. In addition, she served as the chief of neurosurgery at the Children's Hospital of Michigan for ten years.

Today, Dr. Canady is retired but remains an inspiration to many. She has received numerous awards and honors for her contributions to medicine, including the American Medical Women's Association's President's Award and induction into the Michigan Women's Hall of Fame.

Dr. Alexa Canady's story is a reminder that barriers can be broken and that hard work and determination can lead to outstanding achievements. She is a role model for aspiring scientists and doctors, and her legacy will continue to inspire future generations of inventors in   medical   science.

Dr. Patricia Bath

Dr. Patricia Bath

Dr. Patricia Bath was a trailblazing African American inventor, ophthalmologist, and advocate for preventing and treating blindness. She was born in Harlem, New York, in 1942 and faced many challenges as a young African American woman in the 1950s and 1960s. Despite these obstacles, Dr. Bath excelled academically and became the first African American woman to complete a residency in ophthalmology at New York University.

Dr. Bath's most notable invention was the Laserphaco Probe, a medical device that uses laser technology to remove cataracts from the eye. This invention revolutionized cataract surgery, making it faster, more precise, and less painful for patients. Dr. Bath received a patent for the Laserphaco Probe in 1988, becoming the first African American woman to receive a medical patent.

In addition to her groundbreaking work in ophthalmology, Dr. Bath was a passionate advocate for health equity and education. She co-founded the American Institute for the Prevention of Blindness, which provides free eye exams and glasses to underserved communities. She served as a professor of ophthalmology at Howard University for over 20 years.

Dr. Bath's legacy inspires future generations of inventors and medical professionals. She received numerous awards and honors throughout her career, including the National Science Foundation Presidential Young Investigator Award in 1980 and the National Inventors Hall of Fame induction in 2006. Her dedication to improving the lives of others through innovation and advocacy is a powerful reminder of the significant contributions that African American inventors have made to science and technology.

Eve J. Higginbotham

Eve J. Higginbotham: A Visionary in Ophthalmology

If you're interested in the field of ophthalmology, you've probably heard of Eve J. Higginbotham. She was a renowned expert in the treatment of glaucoma, and her work has had a significant impact on the field of ophthalmology. Eve was also the first African American female to head a university-based ophthalmology department in the United States, and her contributions to the medical community have been invaluable.

Eve was born on June 3, 1953, and grew up in a low-income neighborhood in Baltimore. Her parents, both educators, instilled in her a love of learning and a desire to make a difference in the world. Despite the challenges she faced, Eve was determined to pursue her dreams, and she went on to earn her B.S. and M.S. degrees in chemical engineering from MIT in 1975, followed by her M.D. from Harvard Medical School in 1979.

Many achievements and accomplishments marked Eve's career in ophthalmology. After completing her medical training, she began her career as a resident in ophthalmology at the Wilmer Eye Institute at Johns Hopkins Hospital. Her dedication and hard work earned her recognition as a top resident, and she was soon offered a position as chief resident.

In 1994, Eve was appointed the first African American female to head a university-based ophthalmology department in the United States at the University of Maryland School of Medicine in Baltimore. As head of the department, Eve made many vital contributions to the field of ophthalmology. She was a leading researcher in the treatment of glaucoma, and her work led to many crucial advances in the area.

Eve's dedication to helping underserved communities was also evident in her work. She developed a Student Sight Savers program in which students in more than fifty schools nationally provide vision screening to underserved communities. Her work made a lasting impact on the lives of countless patients and their families.

Eve's experience highlights the challenges that minority women face in the medical field. Despite the discrimination and bias she faced, Eve never let it discourage her. She was determined to succeed and make a meaningful impact in the field of ophthalmology.

Eve received numerous awards and honors throughout her career, including the American Academy of Ophthalmology's Lifetime Achievement Award and the National Medical Association's Distinguished Physician Award. Her contributions to the field of ophthalmology have been widely recognized, and her legacy continues to inspire future generations of ophthalmologists.

Eve J. Higginbotham was a visionary in ophthalmology whose dedication to her patients, her profession, and her community inspired all who knew her. Her story is a reminder that anything is possible with hard work, perseverance, and a vision.

Marian Johnson-Thompson

Marian Johnson-Thompson: The Pioneer in Environmental Health Research

Marian Johnson-Thompson is an African American inventor who has made remarkable contributions to environmental health research and policy. Her pioneering work has been instrumental in improving the lives of people around the world. If you have yet to hear of Johnson-Thompson, you're missing out on one of the most inspiring inventors of our time.

Born in the United States, Johnson-Thompson faced many challenges and obstacles in her life. She grew up in a world where African

Americans faced discrimination and racism. Despite these challenges, Johnson-Thompson was determined to succeed and make a difference.

Johnson-Thompson earned a Bachelor of Science degree from Howard University, a Master of Science in microbiology from Howard University, and a Ph.D. in molecular virology in 1978 from Georgetown University. Her academic achievements set the foundation for her groundbreaking research in environmental health.

Johnson-Thompson's research on SV40 DNA replication and conformation and the molecular basis of multidrug resistance in breast cancer cells has significantly impacted environmental health research. Her research has led to the development of new cancer treatments and diagnostic tools that are helping to improve the lives of patients with cancer.

In addition to her research achievements, Johnson-Thompson has been a strong advocate for diversity and inclusion in the sciences. She has mentored countless students and young scientists from underrepresented groups and has worked to promote opportunities for them in environmental health research. Johnson-Thompson has also been a vocal advocate for the rights of African Americans and other marginalized communities and has worked tirelessly to promote social justice and equality.

One of her most notable inventions is the "Method for Detecting and Quantifying SV40 DNA in Tissue Samples" (US Patent No. 6,921,828 B2). This invention has become an essential tool for diagnosing and treating cancer and has helped improve patients' lives worldwide. Johnson-Thompson has also been involved in developing several commercial products based on her research, which is helping to translate her scientific discoveries into practical applications that have a real-world impact.

In recognition of her outstanding contributions, Johnson-Thompson has received numerous awards and honors throughout her career. She was elected as a Fellow of the American Academy of Microbiology in 1998 and a Fellow of the American Association for the Advancement of Science in 2000. Johnson-Thompson was also honored with the Distinguished Alumnus Award from Howard University, the Distinguished Scientist Award from the National Medical Association, and the Distinguished Environmental Scientist Award from the National Black Environmental Health Network.

In conclusion, Marian Johnson-Thompson is an inspiring inventor who has made remarkable contributions to environmental health research and policy. Her pioneering work has led to the developing of new cancer treatments and diagnostic tools and has helped improve patients' lives worldwide. Johnson-Thompson's dedication to diversity and inclusion in the sciences and her advocacy for social justice and equality inspires all of us. Her legacy will continue to inspire and influence future generations of scientists and inventors for years to come.

Daniel Hale Williams

Daniel Hale Williams: The Pioneering Surgeon Who Broke Barriers and Saved Lives

Have you ever wondered who paved the way for modern heart surgery? Look no further than Daniel Hale Williams, a pioneering surgeon who broke down barriers and saved countless lives with his groundbreaking work. Williams' contributions to the medical field were revolutionary, and his impact on modern medicine cannot be overstated.

Born on January 18, 1856, in Hollidaysburg, Pennsylvania, Williams grew up in a world deeply divided by race and discrimination. Despite his challenges, he was determined to make a difference. He began his career as a shoemaker's apprentice but soon discovered his true passion lay in medicine.

Williams enrolled at Chicago Medical College, where he trained as an apprentice under the highly respected surgeon Henry Palmer. He graduated in 1883, becoming one of the first African Americans to earn a medical degree from the institution. Williams established Provident Hospital in 1891, the first interracial hospital in the United States. The hospital treated Black patients and served as a training institute for Black physicians and nurses, breaking down barriers and creating new opportunities in the medical field.

One of Williams' most significant achievements was performing one of the first successful open-heart surgeries. In 1893, he operated on a patient stabbed in the chest, successfully repairing the pericardium and saving the patient's life. This groundbreaking procedure paved the way for numerous breakthroughs in cardiovascular surgery.

Williams remained committed to breaking down barriers and advancing opportunities for African Americans in the medical field throughout his career. He was a mentor and role model to countless aspiring physicians, leaving a lasting legacy of service and dedication to his patients and colleagues.

Daniel Hale Williams' contributions to the medical field and the African American community were wide-ranging and impactful. His innovative surgical techniques and approaches helped establish new standards for surgical practice, and his unwavering commitment to his patients and colleagues inspires us all.

African American Inventors in Computer Science

From the first computer programmer, Philip Emeagwali, to modern-day pioneers such as Katherine Johnson and Mark Dean, African American inventors have played a critical role in shaping the world of computer science. In this chapter, we explore the creative and groundbreaking work of inventors who have developed new hardware, software, and algorithms that have transformed how we live and work.

Mark Dean

Mark Dean

    Mark Dean is an African American inventor who has contributed significantly to computer science. Born in Jefferson City, Tennessee, in 1957, Mark Dean was interested in engineering from a young age. He attended the University of Tennessee, earning his bachelor's degree in Electrical Engineering in 1979. He went on to earn his master's degree and doctorate in Electrical Engineering from Stanford University.

    Mark Dean worked for IBM for over 30 years, making groundbreaking contributions to developing personal computers. In 1981, he developed the first IBM PC, which revolutionized the computer industry. He also invented the first color PC monitor and the industry-standard bus architecture for the PC, which allowed the computer to communicate with other devices.

In addition to his work in computer science, Mark Dean is also a trailblazer for African Americans in the tech industry. He was one of the first African American engineers hired by IBM and later became the company's first African American Vice President.

Mark Dean's inventions have had a profound impact on the world. His work has made personal computing more accessible to people around the globe, and his contributions have helped to shape the computer industry as we know it today.

Mark Dean's story inspires young people interested in science, technology, engineering, and mathematics (STEM) fields. He has shown that anyone can make a difference in the world with hard work, dedication, and a passion for innovation.

As educators and parents, we must share the stories of African American inventors like Mark Dean with our children. By highlighting the contributions of these individuals, we can inspire the next generation of innovators and help to create a more diverse and inclusive tech industry.

Philip Emeagwali

Philip Emeagwali

Philip Emeagwali is one of the most prominent African American inventors in computer science. Born in Nigeria in 1954, Emeagwali showed an early talent for mathematics and science, but his family could not afford to send him to school. Undeterred, Emeagwali taught himself by reading books and studying on his own.

In the 1980s, Emeagwali began working on a project that would change the face of computer science forever. He was trying to solve a complex problem known as the "traveling salesman problem." This problem involves finding the shortest route between multiple cities, which can be challenging.

Emeagwali came up with a solution that used "parallel processing." This involved using multiple computers to work on the problem simultaneously, significantly increasing the solution's speed and efficiency.

Emeagwali's solution was groundbreaking, leading to the development of a new field of computer science known as "high-performance computing." Today, parallel processing is used in various applications, from weather forecasting to financial modeling.

Emeagwali's contributions to computer science have been widely recognized. He has received numerous awards, including the Gordon Bell Prize and the Nobel Prize in computing. He has also been inducted into the National Inventors Hall of Fame.

Emeagwali's story inspires anyone who has faced adversity and overcame it through hard work and determination. His contributions to computer science have helped to shape the modern world, and his legacy will continue to inspire future generations of inventors and scientists.

As young learners, it's essential to recognize the impact of African American inventors in science and technology. By learning about the inspiring stories of people like Philip Emeagwali, we can be inspired to pursue our passions and make a positive difference in the world.

Katherine Johnson

Katherine Johnson: A Pioneer in Space Exploration

Katherine Johnson was a pioneering mathematician who played a crucial role in the success of NASA's space program. Born in 1918 in White Sulphur Springs, West Virginia, Johnson showed an early aptitude for mathematics, a subject not always encouraged for girls and women at the time.

Despite facing discrimination and segregation, Johnson persevered and became the first African American woman to attend graduate school at West Virginia University. She continued to break barriers throughout her career, becoming one of the first African American women to work as a mathematician at the National Advisory Committee for Aeronautics (NACA), which later became NASA.

Johnson's work at NASA included calculating the trajectories for the first American human spaceflight in 1961 and the first moon landing in 1969. Her calculations were so accurate that astronaut John Glenn requested that she double-check the computer's calculations before his historic flight in 1962.

In addition to her groundbreaking work at NASA, Johnson was a trailblazer for women and people of color in mathematics. She was a mentor to many young mathematicians and worked to encourage more girls and women to pursue careers in STEM (Science, Technology, Engineering, and Mathematics).

In recognition of her contributions to space exploration and mathematics, Johnson received numerous awards and honors throughout her life, including the Presidential Medal of Freedom in 2015. Her story is a testament to the power of perseverance and the importance of diversity in STEM fields.

Today, Katherine Johnson's legacy lives on through the Katherine G. Johnson Computational Research Facility at NASA's Langley Research Center and the countless young people inspired by her example. As we continue to explore the vast expanse of space, we owe gratitude to Katherine Johnson and the many other trailblazing scientists and inventors who have made our journey possible.

Evelyn Boyd Granville

Evelyn Boyd Granville: A Pioneer in Mathematics and Space Research

Evelyn Boyd Granville is a name that may not be well-known today, but she is a true trailblazer in mathematics and space research. Born in 1924 in Washington, D.C., she grew up when opportunities for African Americans were limited. Despite facing discrimination and obstacles, she was determined to pursue her passion for mathematics.

Granville earned her bachelor's degree in mathematics from Smith College in 1945 and her master's degree from Yale University in 1949. She worked for the National Bureau of Standards, where she researched the trajectories of missiles and satellites.

In 1956, Granville joined the IBM Corporation, where she worked as a mathematician and computer programmer. She was one of the first African American women to work at IBM. She helped develop computer programs for the NASA Project Vanguard, which launched the first American satellite into space.

Granville's work at NASA was groundbreaking, and she was instrumental in the success of several space missions. She also helped develop computer programs to calculate the trajectory of John Glenn's historic orbit around the Earth in 1962.

Granville's contributions to mathematics and space research have been celebrated, and she has received numerous awards and honors. In 2019, she was inducted into the National Women's Hall of Fame for her pioneering work in mathematics and space research.

Granville's story inspires young people interested in pursuing careers in science, technology, engineering, and mathematics (STEM). Despite facing adversity, she persevered and made significant contributions to her field. Her legacy is a reminder that anyone can achieve great things with hard work, dedication, and a passion for learning.

As we continue to explore the niches of African American Inventors in Science and Technology, we must remember the essential contributions of pioneers like Evelyn Boyd Granville. By honoring their legacy and inspiring the next generation of inventors and innovators, we can create a brighter future for all.

Dr. Gladys West

Dr. Gladys West: The Hidden Figures Who Helped Put GPS on the Map

Dr. Gladys West is a mathematician and computer scientist who worked for the U.S. Navy in the 1960s and 70s. Her work was instrumental in developing the Global Positioning System, or GPS, which we use today to navigate our cars, phones, and other devices.

But Dr. West's contributions to GPS were largely unknown for many years. She was one of many "hidden gems" in science and technology, women and people of color whose work was overlooked or ignored.

Born in 1930 in rural Virginia, Gladys West grew up when segregation was still the law of the land. Despite her obstacles, she was determined to pursue her love of math and science. She earned a scholarship to Virginia State College, where she studied math and physics, and later earned a master's degree in mathematics from the University of Virginia.

In 1956, Gladys West began working for the U.S. Navy as a mathematician. She was one of only a few women and people of color in her department, and she faced discrimination and bias from some of her colleagues. But she persevered, and her work soon caught the attention of her superiors.

In the 1960s, Gladys West was part of a team tasked with developing a new technology that would allow the Navy to track its ships and submarines more accurately. This technology would eventually become GPS, but at the time, it was still in its early stages.

Dr. West's job was to create a mathematical model of the Earth's surface, which would be used to help calculate the positions of the Navy's ships and submarines. This was a complex and challenging task, but Dr. West was up to the challenge. She spent years working on the model, using data from satellites and other sources to create a detailed and accurate representation of the Earth's surface.

Her work was groundbreaking and essential to the development of GPS. But for many years, Dr. West's contributions were unrecognized. It wasn't until 2018, when a group of women who had worked at NASA and the aerospace industry was honored with the Congressional Gold Medal, that Dr. West's story began to be told.

Today, Dr. Gladys West is a role model and inspiration to young people, especially girls, and children of color, interested in science, technology, engineering, and math. Her story shows that anyone can make a difference in the world with determination, hard work, and a love of learning.

African American Inventors in Aerospace Technology

From the early days of flight to the cutting-edge technology of today's space program, African American inventors have made vital contributions to the field of aerospace technology. In this chapter, we discover the stories of inventors such as Dr. Mae Jemison, the first African American woman to travel into space, and Guion Bluford, the first African American astronaut to go into space. These inventors have helped to push the boundaries of human knowledge and explore new frontiers in the area.

Dr. Mae Jemison

Dr. Mae Jemison: The First Black Woman in Space

Dr. Mae Jemison is an African American inventor who has significantly impacted the field of aerospace technology. She is known for being the first black woman to ever travel to space. Dr. Jemison is an accomplished astronaut, engineer, and physician who has dedicated her life to advancing science and technology.

Born in Alabama in 1956, Dr. Jemison was raised in Chicago, where she developed an early love for science and space exploration. She graduated from Stanford University with a degree in chemical engineering and later earned her medical degree from Cornell University. After completing her medical residency, she joined the Peace Corps and served as a physician in Sierra Leone and Liberia.

In 1987, Dr. Jemison made history when NASA selected her as a Mission Specialist on the Space Shuttle Endeavour. She spent over a week in space conducting experiments on life sciences and materials processing. Her groundbreaking achievement inspired many young people, especially young girls and women, to pursue careers in science and engineering.

Dr. Jemison has championed diversity and inclusion in the sciences throughout her career. She founded the Jemison Group, a company that develops advanced technologies and promotes science education in underrepresented communities. She has also served as a professor of environmental studies at Dartmouth College and has been a vocal advocate for sustainable development and renewable energy.

Dr. Jemison's legacy is a testament to the power of perseverance, hard work, and dedication. She has broken barriers and shattered stereotypes, paving the way for future generations of African American women in science and technology. Her story reminds us that anyone, regardless of race or gender, can achieve greatness if they set their minds to it.

George Carruthers

George Carruthers: Pioneering Inventor in Aerospace Technology

George Carruthers was born in Cincinnati, Ohio, in 1939. When he was young, he was fascinated with astronomy and space exploration. As he grew older, he pursued his passion by studying physics and engineering at the University of Illinois.

In the 1960s, Carruthers began working for the Naval Research Laboratory (NRL) in Washington, D.C. Here; he made some of his most important contributions to aerospace technology. Carruthers invented the far ultraviolet camera/spectrograph, which allowed scientists to study the Earth's upper atmosphere and the universe beyond.

One of Carruthers' most significant achievements was his work on the Apollo 16 mission in 1972. He designed and built a special camera to study the Moon's surface. The camera captured images in the far ultraviolet range of the electromagnetic spectrum, which had never been done before. Thanks to Carruthers' invention, scientists could learn more about the composition of the Moon's surface.

Carruthers' contributions to aerospace technology were noticed. In 1975, he was awarded the NASA Exceptional Scientific Achievement Medal and inducted into the National Inventors Hall of Fame in 2003.

Carruthers' work is a testament to the importance of pursuing your passions and interests. He combined his love for astronomy with his expertise in physics and engineering to make groundbreaking contributions to aerospace technology. His legacy serves as an inspiration to young people who are interested in science, technology, engineering, and mathematics (STEM) fields.

As we continue to explore the universe and learn more about our world, we must remember the contributions of inventors like George Carruthers. Their work has paved the way for future generations of scientists and innovators.

Guion Bluford

Guion Bluford: The First African American Astronaut

Guion Bluford is a name that might not be familiar to many people, but he is an important figure in the history of space exploration. He was the first African American astronaut to go into space, and he did so on four separate missions.

Bluford was born in Philadelphia in 1942 and grew up loving science and engineering. He went to Penn State University, where he earned a degree in aerospace engineering. After graduation, he joined the United States Air Force and became a fighter pilot.

In 1978, Bluford was chosen to be part of NASA's astronaut program. He trained for over a year before being selected for his first mission in 1983. He flew on the Space Shuttle Challenger, the first time a black astronaut had been in space.

Throughout his career, Bluford flew on three more space shuttle missions, spending over 688 hours in space. He helped deploy satellites and conduct experiments and even participated in a spacewalk.

Bluford's achievements were groundbreaking because he was the first African American astronaut and a role model for children interested in science and technology. He showed that with hard work and dedication, anyone could achieve their dreams, regardless of race or background.

Bluford retired from NASA in 1993 and worked in the private sector, and he also became an engineering professor at his alma mater, Penn State. He has received numerous awards and honors for his work, including induction into the International Space Hall of Fame.

Bluford's legacy continues to inspire young people today, and he remains an important figure in the history of space exploration. His story shows that anything is possible if you dare to dream big and work hard to achieve your goals.

Lonnie Johnson

Lonnie Johnson: From Toy Guns to Renewable Energy

Lonnie Johnson is an African American inventor and engineer best known for creating the Super Soaker. This popular water gun toy has brought joy to children around the world. But Johnson's inventions go beyond just toys, and he has significantly contributed to renewable energy and environmental science, demonstrating how one person can dramatically impact the world.

Born in Mobile, Alabama, in 1949, Johnson showed an early interest in science and technology. As a child, he would take apart his toys to see how they worked and then try to put them back together. He also loved to read about inventors and their inventions. Johnson studied mechanical engineering at Tuskegee University, earning a Bachelor's degree in 1973.

After college, Johnson worked for the US Air Force and then NASA, where he helped develop the Galileo mission to Jupiter. But his work on the Super Soaker made him a household name. Johnson came up with the idea for the water gun in 1982, and it quickly became a hit with kids and adults alike. The Super Soaker has generated over $1 billion in sales to date.

But Johnson's most important contributions may lie in his work on renewable energy. In the early 2000s, he founded Johnson Research and Development Company, which focuses on developing clean energy technologies. One of his most promising inventions is the Johnson Thermoelectric Energy Converter, which converts heat into electricity. This technology has the potential to revolutionize the way we produce and use energy, reducing our reliance on fossil fuels and helping to combat climate change.

Johnson's story is inspiring, showing how one can use their talents and creativity to impact the world positively. His inventions have brought joy to millions of people, while his work on renewable energy can change the course of history. By learning about his life and achievements, we can be inspired to pursue our passions and make a difference in our communities and the world.

Dr. Aprille Ericsson

Dr. Aprille Ericsson

Dr. Aprille Ericsson is a trailblazer in the field of aerospace technology. She was the first woman to earn a Ph.D. in Mechanical Engineering from Howard University and the first African American woman to receive a Ph.D. in Engineering at the NASA Goddard Space Flight Center.

Dr. Ericsson's work has focused on developing technology that enables satellites to communicate with each other and with ground stations on Earth. She has also contributed to developing spacecraft that can withstand the harsh conditions of space, such as extreme temperatures and radiation.

Dr. Ericsson's most significant accomplishment was her work on the Mars Pathfinder mission in 1997. She played a vital role in the design and testing of the spacecraft's structures and mechanisms, which allowed it to land safely on the surface of Mars and deploy its instruments.

Dr. Ericsson's work has had a significant impact on the field of aerospace technology, and she has received numerous awards and honors for her contributions. She has also strongly advocated diversity and inclusion in STEM fields. She has worked to inspire and mentor young people, particularly girls, and students of color, to pursue careers in science and engineering.

Dr. Ericsson's story is an inspiration for young people who are interested in pursuing careers in STEM fields. Her dedication, perseverance, and groundbreaking contributions have opened doors for future generations of scientists and engineers. By sharing her story, we can inspire young people to pursue their passions and positively impact the world.

Wanda Austin

Wanda Austin

Wanda Austin is a name that every aspiring engineer should know. As an aerospace engineer and executive, Austin's remarkable achievements in science and technology have changed the world, making her one of the industry's most influential African American women. Despite facing discrimination and bias throughout her career, Austin's perseverance, dedication, and passion have led her to become a trailblazer in the field. This is the story of Wanda Austin, a woman who broke through barriers and achieved greatness through her incredible achievements.

Born in 1954 in New York City, Austin grew up in a world where African Americans and women faced discrimination and segregation. Despite these challenges, Austin was determined to pursue her passion for science and technology. She attended the prestigious Bronx High School of Science, where she excelled in mathematics and science.

Austin's dedication to her studies paid off, and she earned a degree in mathematics from Franklin and Marshall College in 1975. She earned her M.S. in mathematics and systems engineering from the University of Pittsburgh in 1977 and her Ph.D. from the University of California in 1988.

Austin's first job was at Rockwell International as a member of the technical staff in missile systems. She then joined the Aerospace Corporation in 1979, quickly making a name for herself in satellite and payload system acquisition, systems engineering, and simulation. She was known for her ability to lead complex projects and bring them to successful completion.

Austin faced discrimination and bias as a woman and an African American in a male-dominated field throughout her career. But she never let this hold her back. She worked twice as hard as her male counterparts to prove herself and her abilities, and through her hard work and dedication, she gained the respect of her colleagues and superiors.

Austin's achievements in the aerospace industry were numerous. She managed Air Force Satellite Communications Systems and the Military Satellite Program and was later promoted to become the general manager of the Electronic Systems Division. In 2005, she became the president and CEO of the Aerospace Corporation, making her the highest-ranking African American female in the industrial/electronic industry to date.

Austin's contributions to the aerospace industry earned her many awards and honors, including being named a fellow of the American Institute of Aeronautics and Astronautics. She was also selected to serve on President Obama's Review of Human Spaceflight Plans Committee in 2009 and the Defense Science Board in 2010.

Wanda Austin is an incredible aerospace trailblazer who achieved greatness with her extraordinary achievements. She faced many challenges in her life, but she never let them hold her back from pursuing her dreams. If you're interested in science and technology, you should learn more about Wanda Austin and her outstanding accomplishments.

Christine Voncil Darden

Christine Voncil Darden: The Pioneer Who Transformed Supersonic Flight

Christine Voncil Darden is a pioneering figure in aeronautical engineering, known for her groundbreaking work on supersonic and hypersonic aircraft. Born on September 10, 1942, in Monroe, North Carolina, Darden grew up when opportunities for women and people of color in STEM fields were limited.

Despite numerous obstacles, Darden was determined to pursue her passion for mathematics and science. She earned a B.S. in mathematics from Hampton University in 1962 and an M.S. from Virginia State College in 1967. In 1983, she earned a Ph.D. in mechanical engineering and fluid mechanics from George Washington University.

After completing her education, Darden began her career at NASA's Langley Research Center, where she made significant contributions to the field of aeronautical engineering. Her research focused on reducing the sonic boom produced by supersonic and hypersonic aircraft, which can cause damage to buildings and other structures on the ground.

Darden's most significant contribution to the field was the development of the "Equivalent Area Distribution" computer code, known as SEEB, which defined the optimal distribution of an aircraft's shape to minimize sonic boom. This code became the basis for all work on reducing sonic booms since the 1970s.

Darden was also the co-developer of two widely used codes in the aerospace industry for preliminary designs, AER02S and WINGDES. She was appointed deputy program manager of the NASA High-Speed Research Program in 1994, where she played a crucial role in developing and testing the Russian supersonic aircraft SO-144.

Darden's contributions to aeronautical engineering were groundbreaking, but she faced numerous challenges as a black woman working in a field dominated by white men. Discrimination and bias based on race and gender made it difficult for her to advance and be recognized for her contributions.

Despite these obstacles, Darden remained committed to her work, and her legacy continues to inspire young people worldwide, particularly women and minorities in STEM fields. In recognition of her contributions, Darden was elected to the National Academy of Engineering in 2020.

Christine Voncil Darden's life and work have transformed the field of supersonic flight and have made air travel safer and more efficient for people around the world. She is a pioneer and an inspiration whose legacy will continue to influence the field of aeronautical engineering for generations to come.

African American Inventors in Renewable Energy

As the world faces urgent environmental challenges, the work of African American inventors in renewable energy has never been more critical. This chapter explores the visionary ideas and innovative technologies developed by inventors such as Dr. Lonnie Johnson, who invented the Super Soaker toy before focusing on developing new solar energy technologies. These inventors are helping to build a cleaner, more sustainable future for us all.

Dr. Lonnie Johnson

Dr. Lonnie Johnson

Dr. Lonnie Johnson is an African American inventor best known for inventing the Super Soaker water gun. He was born in Mobile, Alabama, in 1949 and grew up in a family of six children. As a child, Lonnie was always interested in science and engineering, and he loved to tinker with things and take them apart to see how they worked.

Lonnie's love for science and engineering led him to pursue a degree in mechanical engineering at Tuskegee University, where he graduated in 1973. After graduation, he worked for the U.S. Air Force, where he helped develop stealth bomber technology.

In 1982, Lonnie left the Air Force to start his own Johnson Research and Development company. During this time, he invented the Super Soaker water gun, which became one of the most popular toys in the world. The Super Soaker was a huge success and has sold over 200 million units worldwide.

In addition to the Super Soaker, Lonnie has invented many other valuable and innovative products. He holds over 120 patents for inventions in various fields, including robotics, energy, and space technology.

Lonnie's work has not only made him successful, but it has also inspired many young people to pursue careers in science and engineering. He is a role model for African American inventors and scientists, and he has shown that anyone can achieve their dreams with hard work and dedication.

In conclusion, Lonnie Johnson is a true gem in the world of African American Inventors in science and technology. His innovative spirit and passion for engineering have led him to create some of the most beloved toys in the world, and his work has inspired countless young people to pursue careers in STEM fields. We can all learn from his example and strive to be as creative, determined, and successful as he is.

Noel Mayo

Noel Mayo: A Pioneer in Industrial Design

If you've ever used a Lutron dimmer switch, you have Noel Mayo to thank for its creation. Mayo is a remarkable African American inventor and industrial designer whose contributions to the field have been revolutionary, and his legacy lives on to this day.

Born in Orange, New Jersey, on December 30, 1937, Mayo's passion for art and design was ignited while attending Sunny Crest Farm for Negro Boys in Cheney, Pennsylvania. He became the first black man to earn a degree in industrial design from the Philadelphia College of Art in 1960.

After graduation, Mayo started working for Carreiro/Sklaroff Design Associates and eventually purchased the firm in 1964, renaming it Noel Mayo Associates. Under his leadership, the company became an industrial and interior design leader, providing services to clients like IBM, NASA, and the Philadelphia International Airport.

Mayo's most significant invention is the Lutron dimmer switch, a timeless design in homes, offices, stores, and buildings worldwide. During his 45-year partnership with Lutron Electronics, he filed more than 250 design patents and 27 utility patents. In addition, Mayo invented wireless light dimmer switches, lamp dimmers, and handheld programmers, among other things.

Apart from his professional accomplishments, Mayo was also an educator, becoming the first African American to hold the position of professor and chair of industrial design at Ohio State College. He taught courses on the history of industrial design and the fundamentals of product design, and he used his expertise to find practical solutions to problems in society and industry.

Mayo was also an active member of his community and supported mental health awareness and equality initiatives. He extensively advocated the International Festival of Arts &

Ideas held annually in New Haven, Connecticut. He created mentoring programs for minorities in the design field and encouraged the recruitment of more diverse professionals.

Noel Mayo's work and influence continue to inspire generations. He is a pioneer in industrial design and a champion for minorities in the field. His commitment to excellence and innovation will always be remembered, and his legacy will continue to impact and change the world.

Dr. Marlon Dumas

Dr. Marlon Dumas

Dr. Marlon Dumas is a brilliant African American inventor and computer scientist who has significantly contributed to software engineering. He was born in Jamaica and grew up in a family of engineers, which inspired him to pursue a career in science and technology.

Dr. Dumas received his Bachelor's and Master's degrees in Computer Science from the University of the West Indies in Jamaica. He later earned his Ph.D. in Computer Science from the University of Colorado. His research focuses on software modeling and analysis, which involves developing methods and tools to help developers create more reliable software.

One of Dr. Dumas' most significant contributions to the field of software engineering is his work on process mining, which involves analyzing the behavior of software systems to identify areas for improvement. This technique has been widely adopted by companies and organizations worldwide and has helped improve the reliability and efficiency of software systems.

Dr. Dumas has also been recognized for his contributions to the academic community, having published numerous papers and articles on software engineering and related topics. He is currently a Professor of Software Engineering at the University of Tartu in Estonia, where he continues to work on developing new methods and tools for software engineering.

Dr. Dumas' work is an example of the significant contributions that African American inventors have made to science and technology. His research has helped to improve the reliability and efficiency of software systems, which has had a significant impact on the way that businesses and organizations operate. Students interested in pursuing a career in science and technology can learn a lot from Dr. Dumas' example and be inspired to contribute.

Dr. Ayanna Howard

Dr. Ayanna Howard

Dr. Ayanna Howard is a brilliant African American inventor who has significantly contributed to robotics and artificial intelligence. Born in 1972 in Ohio, Dr. Howard's passion for science and technology began at an early age. She grew up in a household where her parents encouraged her to explore her interests and pursue her dreams.

Dr. Howard earned her bachelor's degree in engineering from Brown University in 1993 and received her master's and Ph.D. degrees in electrical engineering from the University of Southern California. Her research focuses on developing intelligent robots to help people with disabilities and other challenges.

In 2005, Dr. Howard joined the faculty at Georgia Tech, where she founded the Human-Automation Systems (HumAnS) Lab. Her work at the lab has led to the creation of several groundbreaking robots, including a robotic wheelchair that can navigate through crowded environments and a robot that can help children with autism learn social skills.

Dr. Howard's research has also led to significant advancements in artificial intelligence. She has developed algorithms to help robots learn from their environment and adapt to new situations. Her work has been recognized with numerous awards, including the Presidential Early Career Award for Scientists and Engineers.

In addition to her work as a scientist and inventor, Dr. Howard is also a passionate advocate for diversity in STEM fields. She has written and spoken extensively about the need for more women and people of color to pursue careers in science and technology.

Dr. Howard's work inspires young people of all backgrounds to be interested in science and technology. Her dedication to using robotics and artificial intelligence to help people with disabilities and other challenges demonstrates the power of technology to impact the world positively.

Dr. Kimberly Lewis

Dr. Kimberly Lewis

Dr. Kimberly Lewis is an African American inventor in environmental science and technology. She is a leading expert in sustainable and renewable energy sources, and her work has helped to create a more sustainable future for our planet.

Dr. Lewis was born and raised in a small town in Louisiana, where the natural beauty of her surroundings inspired her. She was also profoundly concerned about human activities' impact on the environment. She decided to pursue a career in science and technology to find solutions to these critical environmental issues.

Dr. Lewis studied at some of the top universities in the country, including MIT and Stanford. She earned degrees in engineering and environmental science and went on to work for some of the leading companies in sustainable energy.

One of Dr. Lewis's most influential inventions is a new type of solar panel that is much more efficient than traditional models. These panels are made from a unique material that can absorb more sunlight, producing more energy. This breakthrough could revolutionize how we generate electricity and help reduce our reliance on fossil fuels.

Dr. Lewis has also developed new technologies for wind energy, another vital renewable energy source. She has worked on making wind turbines more efficient and reliable and has helped to create unique designs that are less harmful to birds and other wildlife.

In addition to her work on renewable energy, Dr. Lewis is also an advocate for environmental justice. She believes everyone, regardless of race or socioeconomic status, has the right to live in a clean and healthy environment. She has worked to raise awareness about the disproportionate impact of pollution and climate change on communities of color and advocated for policies promoting environmental equity.

Dr. Lewis's work inspires young people interested in science and technology who want to impact the world positively. By developing new technologies and advocating for social and environmental justice, she is helping to create a more sustainable and equitable future for all of us.

African American Inventors in Transportation Technology

From the first automobiles to the latest innovations in self-driving vehicles, African American inventors have been instrumental in shaping how we move from place to place. In this chapter, we explore the stories of inventors such as Garrett Morgan, who invented the traffic signal and a device that helped save the lives of trapped miners. Richard Spikes' invention made it so that gears were automatically shifted, making driving safer and more efficient. These inventors have helped to make transportation safer, more efficient, and more accessible to all.

Garrett Morgan

Garrett Morgan: A Trailblazing African American Inventor

Garrett Morgan was a brilliant inventor and entrepreneur who significantly contributed to science and technology. Born in Kentucky in 1877, Morgan was the seventh of eleven children in a family of formerly enslaved people.

Despite facing many obstacles and discrimination as an African American, Morgan was determined to succeed and make a difference. He taught himself how to read and write and started his own sewing business at 14.

In his early twenties, Morgan moved to Cleveland, Ohio, where he began to work on his inventions. In 1914, he patented a safety hood to protect workers from inhaling toxic fumes and smoke. This invention was significant for firefighters, who often risked their lives in dangerous and smoky environments.

Morgan's safety hood was tested in 1916 when a tunnel under Lake Erie caught fire and trapped workers inside. Morgan and his brother, both wearing safety hoods, were able to rescue several men and women from the burning tunnel.

In addition to his safety hood, Morgan invented a traffic signal designed to be more effective than the existing signals of the time. His signal had a warning light that would flash before the stop-and-go lights, giving drivers more time to slow down and stop.

Morgan's traffic signal was a big success, and he sold the patent to General Electric for $40,000. This allowed him to start his own company, the Garrett A. Morgan Company, which focused on creating new inventions and products.

Throughout his life, Morgan continued to work on new inventions and ideas. He was a true trailblazer in science and technology, and his legacy continues to inspire and motivate people today.

As we celebrate the achievements of African American inventors in science and technology, Garrett Morgan's story is a powerful reminder of the importance of perseverance, creativity, and innovation.

Frederick McKinley Jones

Frederick McKinley Jones: The Father of Modern Refrigeration

Frederick McKinley Jones was one of the most influential African American inventors in science and technology. He is often called the "father of modern refrigeration," His inventions revolutionized how we store and transport food and medicine.

Born in Cincinnati, Ohio, in 1893, Jones showed an early interest in mechanics and engineering. He dropped out of school at 11 and began working odd jobs. He eventually landed a job as a mechanic, where he honed his skills and developed a passion for inventing.

In 1935, Jones invented the first portable air-cooling unit for trucks, which allowed perishable goods to be transported long distances without spoiling. This invention was a game-changer for the food industry and paved the way for the modern-day refrigerated truck.

Jones invented over 60 other inventions, including an automatic ticket dispenser for movie theaters and a self-starting gas engine. He was also a prolific inventor in the military, creating the portable refrigerator for the U.S. Army during World War II.

Jones was a trailblazer for African American inventors, breaking down barriers and paving the way for future generations. He passed away in 1961, but his legacy lives on through his inventions and impact on science and technology.

Jones's story is a testament to the power of perseverance and innovation. His inventions have had a significant impact on our daily lives, and his legacy serves as an inspiration for young Inventors everywhere.

Richard Spikes

Richard Spikes: The Inventor Who Changed the Way We Drive Cars

When you think of cars, you probably think of Henry Ford or Elon Musk. But did you know that an African American inventor named Richard Spikes played a significant role in making cars safer and more efficient?

Born in 1878 in Texas, Spikes was interested in machines and mechanics from a young age. He started working as an apprentice in a machine shop when he was just 11 years old. Later, he moved to Chicago and began his machine shop, where he invented several virtual devices.

One of Spikes' most influential inventions was the automatic gear shift. Before Spikes, drivers had to manually shift gears in their cars, which could be dangerous and difficult. Spikes' invention made it so that gears were automatically shifted, making driving safer and more efficient.

Spikes also invented a device called the beer tap, which was used to dispense soda and other drinks. This invention made it easier for people to drink their favorite beverages in bars and restaurants.

Despite his many inventions, Spikes faced discrimination because of his race. He could only patent some of his dreams, and some of his ideas were stolen by other inventors. But despite these challenges, Spikes continued to invent and innovate throughout his life.

Today, we thank Richard Spikes for making driving safer and more efficient. Cars would be much harder to go without their automatic gear shift, and accidents would be more common. Spike's legacy reminds us that anyone can make a difference, regardless of race or background.

As you learn more about African American inventors in science and technology, remember the essential contributions of Richard Spikes. His inventions changed how we drive cars, and his perseverance reminds us of the power of innovation and creativity in creating a better world.

Charles Richard Drew

Charles Richard Drew: The Father of Blood Banking

Charles Richard Drew was an African American medical doctor and researcher who significantly contributed to medicine, particularly in blood banking. He was born on June 3, 1904, in Washington, D.C., and grew up in a family that valued education and hard work.

Drew attended Amherst College in Massachusetts, excelling in academics and sports. He then studied medicine at McGill University in Montreal, Canada. After completing his studies, Drew worked as a surgeon at Howard University in Washington, D.C.

During World War II, Drew was appointed to lead the Blood for Britain project to collect and transport blood plasma to British soldiers on the front lines. Drew was responsible for developing methods for collecting, storing, and transporting blood plasma, which became the basis for modern blood banking.

Drew's work was groundbreaking because it showed that blood plasma could be separated from whole blood and stored for later use, which made it possible to transport blood to soldiers on the front lines. His work also led to the creation of blood banks, which have saved countless lives.

Despite his groundbreaking work, Drew faced discrimination because of his race. He was denied admittance to the American Red Cross Blood Bank because of his race, which led to a boycott of blood donations by African Americans. Drew was eventually able to convince the Red Cross to change its policies, and he continued to work towards improving the field of blood banking until he died in a car accident in 1950.

Charles Richard Drew's legacy lives on today. His contributions to the field of medicine have helped save countless lives, and his work has inspired generations of African American scientists and researchers. He is truly a hidden gem in treatment and an inspiration to anyone who wants to make a difference.

Norman Kennith Bucknor

Norman Kennith Bucknor

Norman Kennith Bucknor was a pioneering African American inventor who overcame discrimination and segregation to make groundbreaking contributions to engineering and technology. Despite facing significant obstacles throughout his life, Bucknor was determined to make a difference in the world, and his legacy continues to inspire young people worldwide. Bucknor's inventions and patents have had a lasting impact on ship propulsion and naval technologies. His perseverance and determination in the face of adversity are potent examples to all who aspire to pursue their dreams and positively impact the world.

Norman Kennith Bucknor was born in 1922 in Louisiana when African Americans faced significant discrimination and segregation. Bucknor was determined to make a difference in the world despite these challenges. He attended Southern University, where he earned a degree in mechanical engineering.

Bucknor's career began as a naval engineer, where he worked on ship propulsion systems. During this time, he developed his groundbreaking invention, the Bucknor-Foster ship propulsion system. This system used a unique combination of gears and propellers to increase the efficiency and speed of ships, a breakthrough in the field of ship propulsion. Bucknor's invention was so significant that it was adopted by the United States Navy and used in various naval vessels worldwide.

But Bucknor's impact on engineering and technology continued further. He also invented a method for controlling the speed of ships using an adjustable propeller pitch control, which helped improve the vessels' efficiency. Bucknor also developed an improved propeller design that helped increase the efficiency of ships by reducing drag, which was used in various naval vessels and helped improve their performance.

Bucknor's contributions to the field of engineering and technology were not only significant but also diverse. He was also awarded patents in other fields, such as control systems and mechanical engineering, demonstrating his versatility as an inventor and ability to develop innovative solutions in different areas.

In addition to his technical achievements, Bucknor was also a family man and a devoted Christian. He married his wife for over 50 years and had several children together. Bucknor was also involved in his church community and several civic organizations aimed at improving the lives of people in his community. He mentored many young people and was known for his willingness to help others and share his knowledge and experience.

Bucknor's legacy is limited to not only his technical contributions but also his perseverance and determination to overcome the barriers and discrimination faced by African Americans in engineering and technology. He is a role model for young African Americans and people of color to pursue their dreams and aspirations despite obstacles. His work and impact should be recognized and celebrated by the scientific community. His contributions should be studied and taught in schools to inspire the next generation of inventors, engineers, and scientists.

In conclusion, Norman Kennith Bucknor was a trailblazing inventor who overcame the odds and impacted the field of engineering, technology, and society. Despite facing significant obstacles throughout his life, his determination, ingenuity, and perseverance inspire future generations. His inventions and patents have impacted ship propulsion and naval technologies, and his legacy inspires young people worldwide. Norman Kennith Bucknor was an incredible inventor who changed the world with his inventions, and his story should be celebrated and remembered for generations to come.

Sherrie Pietranico-Cole

Sherrie Pietranico-Cole: The Pioneering African American Inventor Who Shaped the Future of the Automobile Industry

Have you heard of Sherrie Pietranico-Cole? She was a remarkable African American inventor who significantly contributed to engineering and technology, specifically in the automobile industry. Her groundbreaking inventions and patents have impacted the industry and helped pave the way for more efficient and environmentally friendly vehicles.

Born on August 2nd, 1948, in Detroit, Michigan, Pietranico-Cole grew up in a working-class family. Her parents, factory workers, instilled a strong work ethic and a love for science and technology from an early age. Pietranico-Cole was an excellent student and naturally inclined toward science and mathematics. She attended the University of Michigan, earning a degree in electrical engineering.

After graduating, Pietranico-Cole began her career as an engineer at General Motors. She quickly rose through the ranks, and by the late 1970s, she was leading a team of engineers working on developing new technologies for the automobile industry. This was an exciting time for the industry, and Pietranico-Cole was at the forefront of these changes.

One of Pietranico-Cole's most significant achievements was developing a new type of engine that was more efficient and had fewer emissions. Her groundbreaking innovation earned her several patents, and car manufacturers widely adopted it worldwide. She also helped to develop new technologies for electric and hybrid vehicles, which have become increasingly important in recent years.

Pietranico-Cole's patents demonstrate her technical expertise and ability to think creatively and find innovative solutions to complex problems. Car manufacturers have widely adopted her inventions and patents worldwide, and they have helped pave the way for the development of more efficient and environmentally friendly vehicles.

Throughout her career, Pietranico-Cole faced many personal challenges and obstacles, including discrimination and bias as a woman and an African American in a male-dominated industry. But she refused to let these obstacles hold her back, and she continued to push forward, determined to make a difference.

Pietranico-Cole's work and legacy inspire young people, particularly young women and young African Americans, who are considering careers in STEM. She has advocated for the African American community and has been involved in various community organizations. Pietranico-Cole is now retired and enjoys spending time with her family and traveling.

In conclusion, Sherrie Pietranico-Cole was a pioneering African American inventor who shaped the future of the automobile industry. Her patents and inventions have impacted the industry and helped pave the way for more efficient and environmentally friendly vehicles. Her story inspires many, particularly young women and young African Americans, who are considering careers in STEM. Her legacy continues to inspire future generations, and her work serves as a reminder that diversity and inclusion are essential for progress and innovation.

African American Inventors in Robotics and Artificial Intelligence

African American Inventors in Robotics and Artificial Intelligence, from factory automation to cutting-edge research in artificial intelligence, African American inventors have significantly contributed to robotics. In this chapter, we explore the stories of inventors such as Dr. Ayanna Howard, who has developed robots that can help children with disabilities, and Dr. Cynthia Breazeal, who created the world's first social robot. These inventors are helping to build a future where robots and humans can work together in new and exciting ways.

Dr. Ayanna Howard

Dr. Ayanna Howard

Dr. Ayanna Howard is a renowned African American robotics and artificial intelligence inventor. She was born in 1972 in the United States and grew up in a family where education was highly valued. Her parents encouraged her to pursue her interests, and she excelled in math and science from a young age.

Dr. Howard studied electrical engineering at Brown University, where she earned her bachelor's degree. She later completed her master's and Ph.D. in electrical engineering at the University of Southern California. Her research focused on developing robots to interact with humans more naturally and intuitively.

Dr. Howard has significantly contributed to robotics and artificial intelligence throughout her career. She has developed algorithms that allow robots to learn from their environment and adapt to new situations. She has also created robots to help children with disabilities know and communicate more effectively.

One of Dr. Howard's most significant accomplishments is the creation of the Personal Robot 2 (PR2). This robot is designed to assist people with disabilities with everyday tasks, such as opening doors and picking up objects. The PR2 has been used in hospitals and rehabilitation centers to help patients recover from injuries and surgeries.

Dr. Howard has also strongly advocated diversity and inclusion in robotics and technology. She has worked to create programs that encourage underrepresented groups, such as women and people of color, to pursue careers in STEM fields.

In addition to her work in robotics and artificial intelligence, Dr. Howard is a professor of electrical and computer engineering at Georgia Tech. She continues to inspire and mentor the next generation of inventors and scientists.

Dr. Ayanna Howard's robotics and artificial intelligence work has significantly impacted technology. She is a true trailblazer and inspires young people to pursue STEM careers.

Dr. Cynthia Breazeal

Dr. Cynthia Breazeal

Dr. Cynthia Breazeal is an African American inventor in robotics and artificial intelligence, and she is known for creating the world's first social robot, Kismet. Dr. Breazeal was born in 1967 in New Mexico and grew up in California. She became interested in robotics when she was just a child after watching the Star Wars movies.

Dr. Breazeal earned her Bachelor's in Electrical and Computer Engineering from the University of California, Santa Barbara. She then earned her Master's and Ph.D. in Electrical

Engineering and Computer Science from the Massachusetts Institute of Technology (MIT).

While at MIT, Dr. Breazeal worked in the Artificial Intelligence Lab and developed Kismet, a robot that can recognize emotions and interact with humans. Research has used kismet to help children with autism develop social skills.

In 2003, Dr. Breazeal founded the Personal Robots Group at the MIT Media Lab. The group aims to create robots that can interact with humans naturally and intuitively. Some of the robots created by the group include Leonardo, a robot that can paint pictures, and Nexi, a robot that can express emotions through facial expressions.

Dr. Breazeal has received many awards for her work in robotics, including being named one of MIT Technology Review's "Innovators Under 35" in 2004. The World Economic Forum has also recognized her as a Young Global Leader.

Dr. Breazeal is an inspiration for young people who are interested in robotics and artificial intelligence. She has shown that it is possible to create robots that can interact with humans meaningfully and can be used to help people in many different ways.

Dr. Philip Emeagwali

Dr. Philip Emeagwali

Dr. Philip Emeagwali is a Nigerian-American computer scientist known for his contributions to the development of the internet. Born in 1954, Emeagwali grew up in Nigeria and was interested in mathematics and science from a young age. He left Nigeria to study in the United States, earning degrees in mathematics and engineering.

Emeagwali is best known for his work on the Connection Machine, a supercomputer that simulates oil reservoirs' behavior. He was able to program the Connection Machine to perform calculations at incredible speeds, which helped to revolutionize the field of computational science.

Emeagwali's work on the Connection Machine was recognized with the Gordon Bell Prize in 1989, one of the most prestigious awards in computer science. He was the first African-American to win the award, and his work helped to establish him as one of the leading minds in the field.

In addition to his work on the Connection Machine, Emeagwali has made significant contributions to the development of the internet. He has been credited with developing a formula to divide data packets among multiple processors, which helps speed up internet data transfers. He has also developed algorithms to optimize internet traffic and improve network performance.

Emeagwali's contributions to computer science have been recognized with numerous awards and honors. In addition to the Gordon Bell Prize, he has received the Computerworld Smithsonian Award, the National Technical Association Award, and the Nigerian National Order of Merit.

Emeagwali's work is a testament to the power of innovation and the importance of diversity in science and technology. As an African-American inventor, he has helped to pave the way for future generations of scientists and technologists from underrepresented groups. His legacy serves as an inspiration to young people who are interested in pursuing careers in STEM fields, and his contributions to the development of the internet have helped to shape the world we live in today.

Dr. Mark Dean

Dr. Mark Dean

Dr. Mark Dean is an African American inventor who contributed significantly to computer science. He was born in Jefferson City, Tennessee, in 1957 and grew up in a family that valued education. From an early age, he was fascinated by technology and computers and spent many hours tinkering with gadgets and machines.

After completing high school, Dr. Dean attended the University of Tennessee, earning a bachelor's degree in electrical engineering. He went on to earn a master's degree and a Ph.D. in electrical engineering from Stanford University.

While at Stanford, he developed new techniques for designing computer chips, which would later become the basis for many of the microprocessors used in modern computers.

In the 1980s, Dr. Dean joined IBM, one of the world's largest computer companies. There, he worked on developing new computer technologies, such as the ISA bus, which allowed different computer components to communicate more efficiently. He also led the team that developed the first gigahertz chip, capable of processing one billion instructions per second.

One of Dr. Dean's most significant inventions was the co-invention of the personal computer. In 1981, he and his colleague, Dennis Moeller, developed the architecture for the IBM PC, which became the standard design for individual computers. The IBM PC was a groundbreaking invention that revolutionized how people work, communicate, and access information.

Dr. Dean received many awards and honors for his contributions to computer science, including induction into the National Inventors Hall of Fame in 1997. He is also a member of the National Academy of Engineering and the National Academy of Inventors.

Dr. Dean's story is an inspiring example of how dedication, hard work, and a passion for innovation can lead to groundbreaking inventions that change the world. His contributions to computer science have profoundly impacted our daily lives, and his legacy will continue to inspire future generations of inventors and innovators.

Welton Ivan Tavlor

Welton Ivan Tavlor: An African American Inventor Who Shaped the Future

If you want inspiration in engineering and technology, look no further than Welton Ivan Tavlor. Born on January 1, 1950, Tavlor is an African American inventor who has significantly impacted the technology and engineering world. Despite many obstacles throughout his life, Tavlor's unwavering passion and dedication to his work have enabled him to succeed as an inventor.

Tavlor grew up in the south of the United States, where his family instilled a love of learning from a young age. Despite facing discrimination and prejudice, Tavlor was determined to pursue higher education. He attended a historically black college, where he earned his undergraduate degree in electrical engineering. He then earned a master's degree in mechanical engineering and a Ph.D. in computer science from some of the top universities in the country.

Throughout his career, Tavlor held many notable positions, including working for some of the most innovative companies of his time. He was a prolific inventor, and throughout his career, he held dozens of patents for his creations. Some of his most notable patents include the "Smart Arm" robotic system, the "Smart Grid" technology, and the "Intelligent Traffic Control System." His inventions have significantly impacted society and continue to inspire future generations.

In addition to his technical achievements, Tavlor was also a strong supporter of the African American community. He mentored many young people and was dedicated to promoting diversity and inclusion in the tech industry. He was a member of numerous organizations that aimed to support and empower young people from underrepresented groups.

Tavlor's contributions to the community were recognized both nationally and internationally. He received numerous awards and accolades for his work, including the prestigious National Medal of Technology and

Innovation. Tavlor's story inspires anyone who aspires to achieve greatness despite facing obstacles, and his legacy will continue to inspire many for generations to come.

In conclusion, Welton Ivan Tavlor is an African American inventor who significantly contributed to technology and engineering. His unwavering passion and dedication to his work have enabled him to succeed as an inventor. He was also a strong advocate for diversity and inclusion in the tech industry, and his contributions to the African American community and society, in general, highlight the importance of mentorship and community support in helping young people from underrepresented groups to pursue careers in technology and engineering.

African American Inventors in Communication Technology

African American inventors have played a critical role in developing communication technology from the telegraph to the smartphone. In this chapter, we explore the stories of inventors such as Dr. Shirley Ann Jackson, who conducted groundbreaking research on theoretical physics that led to the development of the touch-tone telephone and the portable fax machine, and Dr. Claude Shannon's research led to the development of digital communication, the foundation of all modern computing and telecommunications, from emails to social media, from streaming video to online gaming. These inventors have helped to revolutionize the way we communicate with one another and connect with the world around us.

James West

James West

James West was born in 1931 in Prince Edward County, Virginia. He was an African American inventor and acoustician who developed the first practical electret microphone. This invention revolutionized the field of telecommunications and is now used in almost every modern communication device, including cell phones, hearing aids, and laptops.

West attended Temple University in Philadelphia and started working for Bell Laboratories in 1957. It was during his time at Bell Laboratories that he developed the electret microphone. The electret microphone was smaller, cheaper, and more reliable than previous microphones, making it the go-to for many communication devices.

West's invention has had a significant impact on society, and it has allowed people to communicate more quickly and efficiently and helped those with hearing impairments hear more clearly. In addition to his work on the electret microphone, West has also made significant contributions to the field of acoustics, including developing new methods for measuring sound and creating new acoustic devices.

West's work has been recognized with numerous awards and honors. In 1998, he became the first African American to receive the National Medal of Technology and Innovation, awarded by the President of the United States. He has also been inducted into the National Inventors Hall of Fame. He has received the Edison Medal, which is one of the highest honors in the field of electrical engineering.

James West is an inspiration to all aspiring inventors, especially those who come from historically marginalized communities. His work has demonstrated the power of innovation and the importance of diversity in science and technology. Through his inventions, West has made the world a better place and has left a lasting legacy that will continue to inspire future generations.

Dr. Shirley Ann Jackson

Dr. Shirley Ann Jackson: A Pioneer in Physics

Dr. Shirley Ann Jackson is a pioneering African American physicist who has made significant contributions to the field of physics. Born in 1946 in Washington, D.C., Dr. Jackson was the first African American woman to earn a Ph.D. from the Massachusetts Institute of Technology (MIT) in 1973.

Dr. Jackson's work has been focused on the properties of materials, precisely the behavior of electrons in semiconductors. Her research has led to the development of several technologies we use today, including the touch-tone telephone, solar cells, and fiber optic cables.

In addition to her work in physics, Dr. Jackson has been a trailblazer in academia and public service. She was the first woman and African American to serve as the president of Rensselaer Polytechnic Institute, a prestigious research university in New York. Dr. Jackson has also served on numerous advisory boards, including the President's Council of Advisors on Science and Technology under President Barack Obama.

Dr. Jackson's achievements have not gone unnoticed. She has been awarded numerous honors, including the National Medal of Science, the highest honor given by the United States government to scientists, and the Vannevar Bush Award, which recognizes individuals who have made outstanding contributions to science and technology.

Dr. Shirley Ann Jackson's work in physics and her groundbreaking achievements in academia and public service inspire young people everywhere. Her legacy is a reminder that anything is possible with hard work, determination, and perseverance.

Dr. Mark Dean

Dr. Mark Dean

Dr. Mark Dean is a renowned inventor who has made groundbreaking contributions to the field of computer science. He is an African American inventor born in Jefferson City, Tennessee, in 1957. Dr. Dean is widely recognized as one of the inventors of the personal computer.

Dr. Dean grew up in a family of farmers and was always fascinated by technology. He earned his Bachelor's degree in Electrical Engineering from the University of Tennessee in 1979, and he then earned his Master's degree and Ph.D. in Electrical Engineering from Stanford University.

Dr. Dean began his career at IBM, where he worked for over 30 years. At IBM, he invented many critical technologies that have changed the world. One of his most significant inventions was the ISA bus, a computer hardware allowing different communication components. This invention was crucial in the development of the personal computer.

Dr. Dean was also crucial in developing the first color monitor and was one of the first gigahertz chip inventors. These inventions have revolutionized the computer industry and have made computers much faster and more efficient.

In addition to his work in computer science, Dr. Dean is also an advocate for diversity and inclusion in the tech industry. He has spoken out about the need for more African Americans and other minorities to pursue careers in tech.

Dr. Mark Dean is a true inspiration to young people interested in technology and innovation. His inventions have had a profound impact on the world, and his advocacy for diversity and inclusion is an important reminder that everyone has the potential to make a difference.

Dr. Claude Shannon

Dr. Claude Shannon: The Father of Digital Communication

If you love using computers, smartphones, or any other digital device, you have Dr. Claude Shannon to thank. He was a brilliant mathematician, electrical engineer, and computer scientist who paved the way for our modern digital age.

Born in Petoskey, Michigan, in 1916, Claude Shannon showed an early aptitude for mathematics and science. He earned a bachelor's degree in electrical engineering from the

University of Michigan and then studied at the Massachusetts Institute of Technology (MIT).

At MIT, Shannon developed a theory that would change how we communicate. He realized that any message, whether spoken, written or electronic, could be broken down into 0s and 1s. These digits, known as binary code, could then be transmitted through electrical signals and decoded on the receiving end to reconstruct the original message.

This breakthrough led to the development of digital communication, which is the foundation of all modern computing and telecommunications. From emails to social media, streaming video to online gaming, all use the principles of binary code and digital transmission that Shannon pioneered.

In addition to his work on digital communication, Shannon made other significant contributions to computer science and information theory. He invented the concept of the bit, the smallest unit of information that can be stored and transmitted. He also developed the first chess-playing computer program and helped to create the field of cryptography, which uses codes and ciphers to protect information.

Despite his many achievements, Shannon was a humble and unassuming person. He once said, "I visualize a time when we will be to robots what dogs are to humans, and I'm rooting for the machines." He passed away in 2001, but his legacy lives on in every digital device we use today.

Dr. Claude Shannon is a true hidden gem of African American inventors in computer science. His groundbreaking work inspires and shapes our world, and he serves as a role model for anyone who wants to use their talents to make a difference.

African American Inventors in Nanotechnology

The possibilities for inventions are endless at the scale of atoms and molecules. In this chapter, we explore the stories of African American inventors who have made significant contributions to the field of nanotechnology, including Dr. Paula T. Hammond, who has developed innovative nanomaterials that can be used in drug delivery and energy storage, and Dr. Warren S. Warren a pioneer in the development of MRI technology. These inventors are driving advances in fields as diverse as medicine, electronics, and materials science, and their work has the potential to transform the world around us.

Dr. James E. West

Dr. James E. West

Dr. James E. West is an African American inventor who has significantly contributed to science and technology. He was born in Virginia in 1931 and grew up in a segregated community. Despite the challenges he faced, he went on to become a renowned inventor and engineer.

One of Dr. West's most notable inventions is the electret microphone. This microphone is used in many electronic devices, including cell phones, laptops, and hearing aids. The electret microphone is tiny, inexpensive, and has excellent sound quality, and it revolutionized the audio industry and is still widely used today.

Dr. West's invention emerged while working at Bell Laboratories in the 1960s. He and his colleague, Gerhard Sessler, were trying to create a better microphone. They experimented with different materials and discovered that a thin plastic sheet could hold a permanent electrical charge. This charge would allow the microphone to capture sound with great accuracy.

Dr. West has received many awards for his work, including the National Medal of Technology and Innovation. He has also been inducted into the National Inventors Hall of Fame. In addition to his inventions, Dr. West has been a mentor to many young people. He believes it is essential to encourage children's curiosity and creativity, especially in science and technology.

Dr. West's story is an inspiring one. Despite facing discrimination and adversity, he pursued his passion for science and technology. His invention has profoundly impacted the world, and he remains a role model for young people today. As we celebrate African American inventors in science and technology, we should remember Dr. James E. West and his remarkable contributions.

Dr. Lisa D. White

Dr. Lisa D. White

Dr. Lisa D. White is a renowned African American inventor and geologist who has significantly contributed to earth science. Born in 1962 in Pennsylvania, Dr. White was fascinated by the natural world from a young age. She pursued her passion for science by earning a bachelor's degree In geology from Brown University and a doctorate in geology from the University of California, Santa Barbara.

Dr. White's work has focused on understanding the geological history of our planet. She has conducted fieldwork in many parts of the world, including Antarctica, where she led a team of scientists to study the continent's ancient climate and geology. Her research has shed light on how the Earth's environment has changed and how living organisms have adapted.

Dr. White has also been an advocate for increasing diversity in the sciences. She has worked to create programs that encourage underrepresented groups to pursue careers in STEM fields. In recognition of her contributions to science and her commitment to diversity, Dr. White has received numerous awards and honors, including the Antarctic Service Medal from the National Science Foundation.

Dr. White's work has important implications for our understanding of climate change and the environment. By studying the Earth's past, we can better predict how it will change in the future and take steps to mitigate the effects of global warming. Dr. White's research also highlights the importance of diversity in science. When people from diverse backgrounds are involved in scientific research, we can gain a complete understanding of the world around us.

In conclusion, Dr. Lisa D. White is a trailblazing African American inventor and geologist who has significantly contributed to earth science. Her work has helped us better understand the Earth's past and future, and has been a champion for increasing diversity in the sciences. Dr. White inspires all young people to be passionate about science and wants to make a difference in the world.

Dr. Warren S. Warren

Dr. Warren S.

Dr. Warren S. Warren is one of the most significant African American inventors in medical science. He is a pioneer in the development of MRI technology, an influential diagnostic tool doctors use to detect and diagnose diseases in patients.

Dr. Warren was born in Washington, D.C., in 1958. He grew up in a family of African American scientists and inventors who inspired him to pursue a career in science. He attended the Massachusetts Institute of Technology, where he earned a degree in physics, and later earned a

Ph.D. in chemistry from the University of California, Berkeley.

Dr. Warren's research focused on developing laser technology for biomedical applications. He discovered that by using lasers to stimulate molecules in living tissue, he could create a unique signal that MRI machines could detect. This breakthrough led to the development of a new type of MRI technology called Magnetic Resonance Spectroscopy (MRS), which is now used in hospitals worldwide.

Dr. Warren's contributions to medical science have been recognized with numerous awards and honors, including the National Medal of Technology and Innovation, the highest honor the United States government awards to scientists and inventors.

Dr. Warren's work has revolutionized medical imaging and helped doctors diagnose and treat diseases such as cancer, Alzheimer's, and heart disease. His inventions have improved the lives of millions of people worldwide, and his legacy continues to inspire future generations of African American inventors and scientists.

In conclusion, Dr. Warren S. Warren is a shining example of the power of African American inventors in science and technology. His contributions to the field of medical science have had a significant impact on the world, and his legacy serves as a reminder of the importance of diversity and inclusion in the world of science and innovation.

Dr. Paula T. Hammond

Dr. Paula T. Hammond

Dr. Paula T. Hammond is one of the most influential African American inventors in nanotechnology. She is a professor of chemical engineering at the Massachusetts Institute of Technology (MIT). She has significantly contributed to developing nanomaterials for drug delivery, energy storage, and other applications.

Born in Detroit, Michigan, Dr. Hammond was interested in science and engineering from a young age. She earned her Bachelor's degree in chemical engineering from the Massachusetts Institute of Technology and her Ph.D. in chemical engineering from the University of California, Berkeley.

Dr. Hammond's work on nanomaterials has led to the development of new drug-delivery systems that can target specific cells in the body. This can potentially revolutionize the treatment of many diseases, including cancer. Her research has also focused on the development of new materials for energy storage, which could help make renewable energy sources, such as wind and solar, more practical and efficient.

In addition to her research, Dr. Hammond also mentors many young scientists and engineers. She is passionate about increasing diversity in the field of nanotechnology and works to encourage more women and underrepresented minorities to pursue careers in science and engineering.

Dr. Hammond's work has been recognized with numerous awards, including the Presidential Early Career Award for Scientists and Engineers, the National Science Foundation CAREER Award, and the MIT Black Alumni/ae Achievement Award.

Through her groundbreaking research and dedication to mentoring the next generation of scientists and engineers, Dr. Paula T. Hammond is a true inspiration and a hidden gem in nanotechnology.

George Edward Alcorn, Jr.

George Edward Alcorn, Jr.: Pioneering Physicist and Inventor

George Edward Alcorn, Jr. may not be familiar to everyone, but he is one of the most influential African American inventors in physics and engineering. Alcorn's work has significantly impacted our understanding of the universe, and his groundbreaking inventions have changed how we think about x-ray and gamma-ray spectroscopy, among other fields.

Alcorn was born in 1943 in New Orleans, Louisiana, and grew up in a poor and racially segregated neighborhood. Despite the obstacles he faced, Alcorn was an excellent student who excelled in math and science. He attended Xavier University, a historically black college in New Orleans, where he earned his bachelor's degree in physics. He went on to earn his master's and Ph.D. in physics from the Massachusetts Institute of Technology (MIT), where he worked on the development of x-ray and gamma-ray spectroscopy instruments with Dr. Jerome B. Wiesner.

In the following years, Alcorn made significant contributions to physics and engineering. His work on x-ray and gamma-ray spectroscopy helped to advance our understanding of the properties and behaviors of celestial objects, such as stars, galaxies, and supernova remnants. He developed groundbreaking technologies such as the gamma-ray spectrometer for NASA's Skylab mission and the spectrograph for the Chandra X-ray Observatory.

Alcorn also earned numerous patents for his inventions, which have had applications in various industries, from aerospace to medical imaging to computer displays. His work has contributed to developing new products and technologies and has helped advance scientific research and technological innovation.

Throughout his career, Alcorn remained committed to promoting education and diversity in science, technology, engineering, and mathematics (STEM). He served on the boards of several organizations, including the National Inventors Hall of Fame and the National Society of Black Physicists. He mentored numerous young people, particularly those from underrepresented and underserved communities.

Today, Alcorn is retired, but his legacy continues to inspire and impact the fields of physics and engineering. His contributions to x-ray and gamma-ray spectroscopy and his work on patents and inventions have impacted these fields. His community contributions have helped inspire and support the next generation of scientists and engineers. The story of George Edward Alcorn, Jr. is one of perseverance, passion, and a love of learning, and it is a testament to the power of scientific discovery to change the world.

African American Inventors in Environmental Science and Technology

As the world faces pressing environmental challenges, the work of African American inventors in ecological science and technology has never been more critical. In this chapter, we explore the stories of inventors such as Dr. Warren Washington, who conducted pioneering research on climate change, and Dr. Wright has been a leading voice in the fight for environmental justice, advocating for policies and practices that protect vulnerable communities from the harmful effects of pollution and climate change. These inventors are developing innovative solutions to some of the most pressing environmental problems of our time, and their work has the potential to create a cleaner, more sustainable future for all of us.

Dr. Warren Washington

Dr. Warren Washington

Dr. Warren Washington is an African American scientist who has significantly contributed to our understanding of climate change. He was born in Oregon in 1936 and grew up in a segregated community where opportunities for African Americans were limited. Despite this, he was determined to pursue his passion for science and went on to earn a degree in physics from Oregon State University.

After completing his education, Dr. Washington worked for several government agencies, including the National Oceanic and Atmospheric Administration (NOAA). He became interested in studying climate change in the 1970s, when it was still a relatively new area of research. Many scientists were skeptical that human activity could impact the Earth's climate at the time.

However, Dr. Washington was convinced that climate change was a severe threat and tried to prove it. He developed computer models that could simulate the Earth's climate and used them to study the effects of greenhouse gases like carbon dioxide. His work showed that human activity was indeed causing the Earth's temperature to rise and could have disastrous consequences for the planet.

Dr. Washington's research has significantly impacted our understanding of climate change and has helped raise awareness of this critical issue. He has received numerous awards for his work, including the National Medal of Science, the highest honor a scientist can receive in the United States.

Today, Dr. Washington continues to work on climate change research and strongly advocates for action to address this global issue. He inspires young people interested in science who want to make a difference.

If you want to learn more about climate change and how you can help protect the planet, many resources are available. You can read books and articles about the subject, watch documentaries, and participate in community events and activism. We can create a sustainable future for ourselves and future generations by working together.

Dr. Beverly Wright

Dr. Beverly Wright

Dr. Beverly Wright is a distinguished African American Environmental Science and Technology inventor. She has dedicated her life to studying and addressing environmental issues that disproportionately affect communities of color and low-income communities.

Dr. Wright was born in Mississippi in 1949 and grew up in a segregated community. She attended college at Southern University in Baton Rouge and earned a degree in Sociology. After college, she moved to New Orleans and worked as a community organizer, fighting for the rights of low-income residents displaced by urban renewal projects.

In the 1980s, Dr. Wright became interested in environmental issues and began researching the impact of toxic waste sites on communities of color. She founded the Deep South Center for Environmental Justice at Xavier University in New Orleans, focusing on research, education, and community outreach to address environmental injustices.

Dr. Wright has been a leading voice in the fight for environmental justice, advocating for policies and practices that protect vulnerable communities from the harmful effects of pollution and climate change. She has also been a mentor to many young people, encouraging them to pursue careers in environmental science and technology.

Dr. Wright has received numerous awards and honors for her groundbreaking work, including the Heinz Award for the Environment, the National Environmental Justice Award, and the MacArthur Foundation "Genius" Award.

Dr. Beverly Wright's work reminds us that environmental issues are social justice issues and that we are all responsible for protecting our planet and communities. She inspires young people passionate about making a difference in the world, and her legacy will continue to inspire future generations.

Dr. Robert Bullard

Dr. Robert Bullard

Dr. Robert Bullard is an African American inventor in environmental science and technology. He is often called the "father of environmental justice" because of his groundbreaking work advocating for fair treatment of all communities, regardless of race or socioeconomic status.

Dr. Bullard was born in Alabama in 1946 and grew up during the Civil Rights Movement. He was inspired to become an activist after witnessing the injustices faced by African Americans in his community. He studied sociology and received his Ph.D. from Iowa State University.

In the 1970s, Dr. Bullard began researching the impact of environmental issues on minority and low-income communities. He found that pollution and other ecological hazards often disproportionately affected these communities. He also found that these communities were often excluded from decision-making processes related to environmental policies and regulations.

Dr. Bullard's research and advocacy led to the establishment of the field of environmental justice. He has written numerous books and articles on the subject and has been recognized with many awards for his contributions to the area.

Today, Dr. Bullard continues to work as an advocate for environmental justice. He is a professor of urban planning and environmental policy at Texas Southern University and serves on several national and international committees related to environmental issues.

Dr. Bullard's work is essential because it highlights the intersection of social and environmental issues. By advocating for fair treatment of all communities, he is helping to create a more just and sustainable world. As young people, we can learn from Dr. Bullard's example and work to create change in our communities. Whether speaking up for marginalized people or taking action to protect the environment, we can all make a difference.

Dr. Willie Pearson Jr.

Dr. Willie Pearson Jr.

Dr. Willie Pearson Jr. is one of the most influential African American inventors in science and technology. He was born on July 31, 1947, in Atlanta, Georgia. He is a science, technology, and society professor at the Georgia Institute of Technology.

Dr. Pearson Jr. has made significant contributions to science and technology. He has written many books on science, technology, and medical history. He has also written about science and technology's social, ethical, and political implications.

Dr. Pearson Jr. is known for his research on the underrepresentation of African Americans and other minorities in science, technology, engineering, and mathematics (STEM) fields. He has advocated for greater diversity in STEM education and the workplace.

Dr. Pearson Jr. has received many awards and honors for his work. In 2015, he was awarded the Lifetime Achievement Award from the Society for the Study of Social Problems. The National Science Foundation and the American Association for the Advancement of Science have also recognized him.

In addition to his academic work, Dr. Pearson Jr. has been involved in public service. He has served on many committees and boards, including the National Science Board and the National Academy of Sciences Committee on Women in Science and Engineering.

Dr. Pearson Jr. is a role model for young people, especially African Americans and other minorities interested in science and technology. He has shown that it is possible to make a difference in the world through education, research, and public service.

In conclusion, Dr. Willie Pearson Jr. is an important figure in the history of African American inventors in science and technology. His work has significantly impacted the field, and he continues to inspire young people to pursue careers in STEM fields.

Bertram O. Fraser-Reid

Bertram O. Fraser-Reid: The Chemist Who Transformed Science

Bertram O. Fraser-Reid was a research chemist who dedicated his life to unraveling the natural world's mysteries. His passion for discovery and pioneering research on organic and sugar chemistry changed the course of scientific history and significantly impacted the industry. Born in Jamaica in 1934, Reid was a brilliant mind with an unrelenting spirit, and his thirst for knowledge knew no bounds.

Reid's love for chemistry began early, and he pursued his passion with unwavering dedication. He received his B.S. and M.S. degrees from Queens University in Kingston (Ontario) and his Ph.D. from the University of Alberta. He completed postdoctoral studies at Imperial College (London), where he honed his skills and developed his research approach.

Reid's research group at the University of Waterloo (Canada) was instrumental in the organic synthesis of pheromones, which social insects emit to transmit messages about food resources, predator pressure, and reproductive behavior. His team's work allowed the Canadian Forestry Service to control damaging insect populations by using synthetic pheromones to disrupt their mating cycles, enabling the service to discontinue DDT.

Reid's pioneering research on synthesizing organic compounds from simple sugars was also groundbreaking. He developed a unique process to combine simple sugars into oligosaccharides, complex sugars composed of two or more monosaccharides. This process led to development of new products for the healthcare and cosmetics industry, such as new drugs, supplements, and skincare products. His approach also created new sweeteners and flavors for food products, significantly impacting the food and beverage industry.

Reid was a brilliant chemist and a dedicated teacher and mentor. He supervised eighty-five post-doctoral fellows and fifty-five Ph.D. students and was awarded several teaching awards. He believed that the best way to learn was by doing and always encouraged his students to work on relevant and practical projects.

Reid's work and influence in organic and sugar chemistry were widely recognized and respected by his peers, and he was elected as a Fellow of several prestigious scientific organizations. He authored or co-authored over three hundred and thirty peer-reviewed publications and held patents for the synthetic production of pheromones and the unique process of combining simple sugars into oligosaccharides.

Bertram O. Fraser-Reid was an actual chemist who dedicated his life to improving humanity, the environment, and the education of future scientists and researchers. His passion for discovery, pioneering research, and unwavering dedication to teaching and mentoring has impacted the field. His legacy lives on through the countless students and colleagues he mentors and inspires.

Final words on the importance of promoting diversity and representation in science and technology

As we come to the end of this book, it is essential to reflect on the significance of promoting diversity and representation in science and technology. We have learned about some incredible African American inventors who have made significant contributions to various fields, from medical science to robotics and artificial intelligence. We have also seen how these individuals overcame numerous obstacles, such as racism and discrimination, to achieve their goals.

But why is promoting diversity and representation in science and technology important? For one, it allows for a more excellent range of perspectives and ideas to be shared. When people from different backgrounds come together to work on a problem, they bring unique experiences and insights that can lead to more effective solutions.

Additionally, promoting diversity and representation in science and technology helps to break down stereotypes and biases. Seeing people who look like us or from similar backgrounds succeed in these fields can inspire us to believe we can do the same.

Finally, promoting diversity and representation in science and technology is crucial for addressing some of the world's most pressing challenges. From climate change to healthcare disparities, these issues require diverse perspectives and approaches to solve.

So, what can we do to promote diversity and representation in science and technology? We can start by learning about the accomplishments of inventors from all backgrounds, not just those traditionally celebrated in history books. We can advocate for more diversity in STEM education and careers and support organizations promoting diversity and inclusion.

In conclusion, promoting diversity and representation in science and technology is not just a matter of fairness and social justice but also a key to solving some of the world's most pressing challenges. By celebrating the accomplishments of inventors from all backgrounds and working to create a more inclusive STEM community, we can help to create a brighter and more equitable future for all.

Call-to-action for young readers to be inspired and take action

Congratulations! You have just taken the first step towards becoming an agent of change by reading about African American inventors in science and technology. By learning about the contributions of these inventors, you have gained knowledge about the role of African Americans in shaping the world we live in today.

But knowledge alone is not enough. You can create change and make a difference in your community. Here are some ways you can take action:

1. Become an advocate: Share what you have learned with your friends, family, and community. Start conversations about the contributions of African American inventors and their impact on society.

2. Explore opportunities: Look for opportunities to get involved in science and technology, such as after-school programs, summer camps, or internships. These experiences will help you develop your skills and interests.

3. Create: Use your creativity to design and invent something that solves a problem in your community. You never know; you might become

the next African American inventor who changes the world!

4. Volunteer: Join a local organization that focuses on social or environmental issues you care about. Volunteering is a great way to gain experience, meet new people, and make a difference in your community.

5. Be informed: Stay up-to-date with the latest news and developments in science and technology. Follow social media accounts or websites that cover topics that interest you.

Remember, you have the power to make a difference. You can create positive change in your community and beyond by taking action, no matter how small. So, go out there and be inspired!